SpringerBriefs in Physics

T0192198

For further volumes:
http://www.springer.com/series/8902

Arkady Plotnitsky

Niels Bohr and Complementarity

An Introduction

 Springer

Arkady Plotnitsky
Theory and Cultural Studies Program
Purdue University
West Lafayette, IN
USA

ISSN 2191-5423   ISSN 2191-5431   (electronic)
ISBN 978-1-4614-4516-6   ISBN 978-1-4614-4517-3   (eBook)
DOI 10.1007/978-1-4614-4517-3
Springer New York Heidelberg Dordrecht London

Library of Congress Control Number: 2012940961

Springer is part of Springer Science+Business Media (www.springer.com)

Preface

This book offers a comprehensive introduction to Niels Bohr's interpretation of quantum phenomena and quantum mechanics, based on his concept of "complementarity," his arguably most significant conceptual contribution to physics and philosophy. The term "complementarity" carries three interrelated but distinct meanings in Bohr's writings:

(1) a general philosophical concept;
(2) a set of physical concepts, specific to quantum mechanics (as against both classical physics and relativity)—concepts that are instantiations of the general philosophical concept of complementarity but that also have independent significance as physical concepts; and
(3) a particular interpretation of quantum phenomena and quantum mechanics, in part based on this concept and its instantiations.

In this study, I shall, for the sake of clarity, only use the term "complementarity" as referring to the first two meanings of the term, and just speak of Bohr's interpretation of quantum phenomena and quantum mechanics, or Bohr's interpretation of quantum mechanics (which includes an interpretation of quantum phenomena), or Bohr's interpretation. Using the phrase "Bohr's interpretation" requires, however, the following additional qualification. Bohr's views underwent, sometimes significant, changes, which led him to changing his interpretation, in some cases in conjunction with different instantiations of his general concept of complementarity (this concept itself has remained pretty much the same), or with new concepts, such as and in particular those of "phenomenon" and "atomicity." There is, thus, more than one "Copenhagen interpretation" even in Bohr's case, not to speak of a number of other interpretations associated with this rubric, often quite different from any of Bohr's own interpretations. I shall avoid this overused and sometimes still abused rubric, although more recent commentaries have been more cautious in using it. On the other hand, it is, I think, legitimate to speak, following Werner Heisenberg, of the spirit of Copenhagen, "the Copenhagen spirit of Quantum Theory [*Kopenhagener Geist der Quantentheorie*]," defined by certain features shared by some of these interpretations and especially pronounced in Bohr's interpretation (Heisenberg 1930, p. iv).

There are several reasons to pursue the project undertaken by this book, three of which I would especially like to stress here:

(1) The first is Bohr's contribution to our understanding of quantum phenomena and quantum mechanics, and the shaping impact of his thought on the history of quantum mechanics and its interpretation. The significance of this contribution and the pervasiveness of this impact are hardly in question, although the resistance to Bohr's ideas has been considerable, beginning with Albert Einstein, who debated the foundations of quantum theory with Bohr throughout his life.

(2) Offering a discussion of the Bohr–Einstein debate is the second reason for undertaking the project, with an additional aim of better explaining Bohr's position in this debate, which has not always been adequately addressed by commentators on this debate. Einstein's position has fared much better, in part because it has had a greater appeal to most of these commentators than Bohr's position did.

(3) The third reason is that, as one of the contributing factors to the resistance to Bohr's thinking and writings, there has been much confusion concerning the meaning of complementarity and of Bohr's ideas in general, and hence his interpretation or, again, interpretations of quantum phenomena and quantum mechanics. Thus, Einstein said that, "despite much effort [he has] expended on it," he was "unable to achieve [a] sharp formulation of [Bohr's principle of complementarity]," which is not uncommon (Einstein 1949b, p. 674).

One of the sources of this confusion is the difficult nature of Bohr's writings on complementarity and quantum mechanics, most of which are assembled in the volumes collectively entitled, *The Philosophical Writings of Niels Bohr* (Bohr 1987; Bohr 1995). These writings are sometimes criticized for their vagueness and elusiveness, a view that the present author does not share and that this book aims to dispel. Bohr's writings are, as I shall argue, lucid and precise, although (this would be hard to deny) they could be difficult and are, in some respects, unconventional, and hence understanding them requires additional effort even on the part of experts in foundations of quantum theory.

Their difficulty is in part due to the complex and counterintuitive character of quantum phenomena and quantum mechanics, which may not be that inhibiting in presenting the mathematics of the theory, but which makes a physically or philosophically explanatory exposition difficult. If anything, Bohr's writings are sometimes excessively cautious and, hence, frequently qualified. Bohr, who strongly believed that a clear and unambiguous exposition of the complexity of quantum phenomena and quantum mechanics was possible and aimed to offer it (the very concept of complementarity was designed to help this task), grappled with this complexity throughout his life. The fact that this struggle is sometimes reflected in Bohr's writings contributes to their difficulty. Bohr commented on these difficulties on several occasions, for example, in his 1949 "Discussion with Einstein on Epistemological Problems in Atomic Physics." Bohr spoke there of "the inefficiency of expression which must have made it very difficult to appreciate the trend of [his] argumentation" because of certain epistemological complexities specific to quantum phenomena that his writings struggled to capture and to express, complexities that I shall address in detail in this study (Bohr 1949, p. 61).

Some of the difficulties of Bohr's writings are, however, due to their particular character, especially insofar as Bohr's key terms, such as "complementarity," "phenomena," or "atomicity" have unconventional meanings. Although this and several other aspects of Bohr's writings are not uncommon in philosophical discourse, they appear to demand more substantial interpretive effort than is customary in scientific texts. Reading Bohr's works requires paying special attention to particular formulations, carefully adhering to the particular meanings of his terms, understanding the philosophical (rather than only physical and mathematical) structure of his concepts, negotiating the different languages involved, and so forth. These demands have diminished these writings' appeal to many physicists, although there have been notable exceptions, especially among those who were close to Bohr, such as Heisenberg and Pauli.

Only a minimal reliance on mathematics in Bohr's arguments did not help either. This reliance characterizes the practice of theoretical physics and the very way of thinking of most physicists and philosophers of quantum theory. The majority of books and articles in the philosophy of quantum mechanics are marked by a strong presence of technical mathematics, thus in a manifest contrast to Bohr's writings. Bohr's writings are sometimes taken to task for their insufficient attention to the mathematics of quantum mechanics, which, so it is claimed, inhibits the effectiveness of Bohr's argumentation or even makes it inadequate. This criticism is, as I shall argue, not justified either and is often based on misunderstanding. The fact that the technical use of formalism is sparse in Bohr's writings does not mean that Bohr did not give sufficient attention to the *role* of mathematics in quantum theory; quite the contrary.

It is not of course that one cannot disagree with Bohr's views or criticize his arguments, which the present study will do on several occasions. The point instead is that not paying proper attention to the aspects of Bohr's writings indicated above prevents one from an effective reading or, when necessary, meaningful criticism of his arguments. Part of my aim in this study is to take these factors into account as much as possible in order to convey the physical and philosophical content of Bohr's views and ideas more effectively.

This aim notwithstanding, this study can only offer *an* interpretation, one of several possible interpretations, of Bohr's argumentation and of his interpretation, or again, his several interpretations, of quantum phenomena and quantum mechanics. Nothing else is possible. Carl Friedrich von Weizsäcker remarked that Bohr's interpretation "needs an interpretation itself, and only that will be its defense" (Weizsäcker1971, p. 25). The first part of the sentence is self-evidently true. An interpretation of quantum phenomena and quantum mechanics always involves an *interpretation* of this interpretation. A given interpretation only functions or even exists when somebody interprets it, especially beyond its existence and functioning for its author, although ultimately this is true even in this case.

As will be seen later in this study, Bohr reinterpreted his earlier interpretations on later occasions, even if without expressly acknowledging or perhaps realizing that he was doing so. I am not saying that Weizsäcker was unaware of this general situation: he might have been or, in any event, would have probably acknowledged the pertinence of these considerations. His contention instead appears to be that there is a degree of incompleteness to Bohr's argument for his interpretation, and thus that this argument needs to be supplemented (consistently with what Bohr expressly says) by further elaborations, explanations, argumentations, and so forth. As any scholarly study of Bohr, this study will provide this type of supplementation as well. However, I do not think that Weizsäcker's contention and, hence, his claim that Bohr's interpretation requires a special defense is true, at least when it comes to Bohr's ultimate interpretation, discussed in Chap. 9, when Bohr revises and corrects his earlier interpretations, of which there are, again, several, the circumstance that Weizsäcker does not address. At least, Weizsäcker's contention and claim are, again, no more true in the case of Bohr's interpretations than those of most other interpretations, except for the initial version of his interpretation, offered in the so-called Como lecture of 1927. As I shall argue in Chap. 4, some parts of this version are difficult to defend, and Bohr significantly modified his interpretation shortly thereafter. Bohr was not God. He struggled hard with the complexities of quantum phenomena and quantum mechanics, and he revised and changed his views, and made mistakes. Nor, as he acknowledged, could he have fully controlled his texts, hard and even obsessively as he tried to do so. Nobody could achieve this either, especially, again, in dealing with subjects of that magnitude of difficulty.

In view of the circumstances just outlined, the task of making Bohr's thinking and work more broadly accessible may not be easy. It is not insurmountable, however, and this book aims to pursue this task. While aware of the difficulties his own writings may present, Bohr insisted that it is possible to make what is fundamentally at stake in quantum physics available to a willing and open-minded layperson, and this book aims to do so as well. There are a few technical mathematical details in this book, but they are minimal and can be skipped without missing the main points of the elaborations where they occur.

This book is a *philosophical* introduction to Bohr. However, it represents a different aspect or even a different *form* of a philosophy of physics, vis-à-vis most forms of the institutional philosophy of physics and specifically the philosophy of quantum theory, or even most philosophical studies of Bohr. The difference is reflected in my emphasis, beginning with this Preface, on Bohr's *thinking* about quantum mechanics or complementarity. Physics is a product of human thought under complex material (nature), technological (our experimental capacities), psychological, historical, and sociological or cultural conditions. Accordingly, one can pursue a philosophy of physics that attempts to understand how physicists think under these conditions, especially at the time of, and in the process of, making, new discoveries.

The present approach is, thus, different from dealing, as is more common, especially, again, in the institutional philosophy of physics, with the logical-axiomatic structure of quantum theory or with broader epistemological or ontological questions, such as reality and causality. These questions were of course a major part of Bohr's thinking, too, and will be considered here, but considered as part of his *thinking*, which, I shall argue, approaches these questions in a new way, demanded, Bohr argued, by the new situation of physics brought about by quantum phenomena. Physics is thinking, thinking about nature or the ontology of nature— of how nature is. It is thinking about what is true or probable about nature or those aspects of nature that physics could consider. In all modern, post-Galilean physics, classical or quantum (or relativity), this truth or probability is determined by means of mathematics, that is, by means of mathematical theories connected, descriptively or predictively, to the suitably configured numerical data obtained in experiments. How close we come to the ultimate constitution of nature in this way may depend on a given theory or, possibly, on nature itself, that is, on how far nature would allow our mathematical theories and experimental technologies to reach in this regard. It is this last question that is ultimately at stake in the Bohr– Einstein debate on, to return to Bohr's title phrase of his account of this debate, "epistemological questions in atomic physics" (Bohr 1949, p. 32).

A physicist can also explore the *nature* of thinking in physics, sometimes simultaneously with creating new physics, and in this sense become a philosopher of physics in the present sense, which tends to happen at the time of crisis, as it happened in Bohr's work. First, creating a proper theory of quantum phenomena and, second, offering a proper interpretation of both quantum phenomena and quantum mechanics presented exceptionally difficult problems for physics, which led to a crisis. The first problem was essentially solved by Heisenberg's and Schrödinger's invention of quantum mechanics in its respectively matrix and wave form in 1925 and 1926. From that point on, Bohr's thinking, which had confronted this crisis for over a decade by then, was directed toward the second problem, which the concept of complementarity helped him to address and ultimately to solve, or at least to reach as far as it was possible for him in solving it, although this took a while (roughly another decade). For some, especially those whose view is closer to that of Einstein's and who are not satisfied with the kind of solution of this problem Bohr offered, the second and, for some, beginning, again, with Einstein, even the first problem has not yet been solved and, hence, the crisis, at least the epistemological crisis, of quantum physics is still ongoing. The debate concerning quantum phenomena and quantum mechanics continues with undi- minished intensity, and given the irreconcilable nature of opposing philosophical positions, such as those of Bohr and Einstein, this debate may be interminable. This situation, however, by no means inhibits the project of philosophy of physics defined by the exploration of the nature of thinking in physics.

This study is an attempt at this type of philosophy of physics, as specifically applied to Bohr's thinking, which continues, just as does Einstein's thinking, to shape this debate. Given the scope of this study, I will not be concerned with psychological and sociological aspects of Bohr's thinking or quantum-theoretical thinking in general. On the other hand, history will play a significant role in my argument. History is unavoidable in physical thinking, which always builds on preceding thinking in physics, even at the time of new discoveries, however revolutionary or historically unexpected the latter may be. Every physical idea has a history, some trajectories of which may be short and others quite long, sometimes, as in the case of the idea of motion, extending to ancient thought. Thus, Bohr's 1913 atomic theory is inconceivable apart from Planck's and Einstein's earlier work. Heisenberg's thinking concerning quantum mechanics was significantly indebted to Bohr's 1913 atomic theory, and conversely, Heisenberg's thinking concerning quantum mechanics, or that of Schrödinger and Dirac, was crucial to Bohr's thinking. There is also a long history of philosophical thinking behind this thinking, extending, again, to ancient thought for example, the atomism of Democritus and Lucretius. More modern philosophical trajectories extend from the critique of causality in Hume and Kant, and traverse via the thought of Kierkegaard, Nietzsche, and William James, among others. Bernhard Riemann's theory of the functions of complex variables was mentioned by Bohr as one of the earlier sources of complementarity, and, as will be seen in Chap. 10, it might have been one of the most important ones. It would be impossible to explore most of this history, but it is equally impossible to bypass it here or in any earnest attempt at understanding Bohr's thinking, which embodies this history, as does our thinking concerning quantum theory, at this point with this thinking itself as part of this history.

A qualification is in order, however. When I speak of the thinking of Bohr or any given author, such as Einstein or Heisenberg, I do not claim to have a determinable access to this thinking. Such an access is limited even when the author is alive and could, in principle, provide one with as much information as possible concerning this thinking. Instead, I refer to thinking that one can follow and can engage on one's own accord on the basis of certain statements or works of a given author, and then in a particular reading or interpretation, since these works can be interpreted differently, and, in the process, related to different ways of thinking. A proper name, such as Bohr, Einstein, or Heisenberg, is the signature underneath a given work or set of works, a signature that attests to one's role as a creator of these works, which serve as a guidance for the thinking that we can pursue as a result of reading them.

In the process, one can also gain insights into how a given author might actually have thought, but claims to that effect are, again, hard to make with certainty, although they may be probable and even highly probable. Such claims are, I would contend, not necessarily less certain when they are made, as for the most part they will be here, on the basis of published works or available manuscripts presenting their authors' arguments in detail rather than from other evidence—letters, notes, diaries, recorded conversations, etc. Historical scholarship tends to rely more on such sources as providing a more reliable evidence concerning or insights (in particular, eureka-like insights) into the thinking of a given figure, often forgetting that this type of evidence is also comprised of texts that require reading and interpretation, sometimes as much as do published works. Such sources will be used by this study as well, which will, however, exercise some extra caution concerning Bohr's reported statements, sometimes unduly relied upon by commentators. I shall primarily focus, however, on Bohr's published essays or more extended unpublished manuscripts, which are, I would contend, reliable sources for understanding his thinking, in part because, as is well known, Bohr paid a particularly careful attention to his published work, albeit never quite finished to his satisfaction. The very concept of a finished manuscript was alien to Bohr, who saw any of his manuscripts, even already published, as an occasion for further thinking, a set of new questions, rather than of conclusive answers. In any event, even apart from their easier availability, Bohr's published works are especially suited for my main concern in this study: what Bohr's thinking, as manifested in his available texts does for *our thinking*, also when our thinking aims to reach beyond Bohr's own.

An interpretation of quantum phenomena and quantum mechanics only becomes effective or, again, operative to begin with when it becomes part of our thinking, and it advances physics, or the philosophy of physics, when it moves this thinking beyond this interpretation, beyond our interpretation of this interpretation. Bohr might be argued to encourage this approach to his writings, an approach that he even adopted himself in dealing with his writings and that his writings invite us to continue. He *reportedly* said that his "every sentence ... must be understood not as an affirmation, but as a question" (Rosenfeld 1967, p. 63). One must, as I said, be cautious concerning such reported statements. However, although not strictly true factually (there are affirmations, or conversely, negations, in Bohr's arguments), this particular statement does correspond to the spirit of Bohr's thinking and writings.

This does not of course mean that, helpful as it might be, Bohr's thinking and writings, or, again, *interpreting* these writings and thus Bohr's interpretations of quantum phenomena and quantum mechanics, is the only path toward a better understanding of them and, especially, advancing this understanding or quantum theory itself. When it comes to advancing physics one's faithfulness or loyalty to Bohr's or anyone's thinking becomes a secondary matter and only counts insofar as it helps this advancement. While my reading of Bohr must and aims to be faithful to his writing as much as possible and while my project is motivated by my belief in the helpfulness of Bohr's thinking for understanding quantum theory and potentially advancing it, this project is not driven by loyalty to Bohr's ideas themselves. At a certain point, our thinking concerning quantum physics will inevitable have to move beyond Bohr, and there is no special reason to assume that it will do so following the path established by Bohr's thinking. An entirely different trajectory, either already in place or yet to be discovered, may be necessary for this task. The project of this book is an introduction to Bohr's thinking concerning quantum phenomena and quantum theory, and it aims to facilitate our understanding of this thinking. This project will succeed most, however, if it will help our thinking to move beyond that of Bohr, either by still following the trajectory of his thinking or, guided by what quantum phenomena and quantum physics tell us, by finding a different trajectory of thinking.

The conception of the book as outlined here makes it different from most previous approaches to Bohr. There are a number of available comprehensive studies of Bohr, a few of which are cited in the special section of the bibliography under the heading, "Suggested Further Readings." Two previous works by the present author offer extensive discussions of Bohr as well, *Reading Bohr: Physics and Philosophy* (2006) and *Epistemologyand Probability: Bohr, Heisenberg, Schrödinger and the Nature of Quantum–Theoretical Thinking* (2009), although the latter study is only partially devoted to Bohr. While I shall rely here on some of the research used in these earlier studies, especially *Epistemology and Probability*, the present book is very different from them, in part by virtue of its conception as an *introduction* to Bohr's thinking concerning quantum mechanics and complementarity. My aim here is a streamlined discussion maximally centered on Bohr's thinking, concepts, and arguments in their historical development.

In part for this reason, my discussion of the works of other authors, even Heisenberg and Einstein, who were especially germane to Bohr's thought, is limited, and my engagement with secondary literature on Bohr is reduced to bare minimum. Bohr did not of course think in a vacuum, nor does the present author, and this study will address the ideas of other key figures, such as Heisenberg and Einstein, and Bohr's responses to these ideas, and will refer and engage with secondary sources whenever necessary. However, these engagements will play only an auxiliary role here, that of helping the exposition of Bohr's thinking and argumentation. As just mentioned, in addition to the works actually cited by this study, the Bibliography contains a section, "Suggested Further Reading," which lists both the primary works of some of the major authors and scholarly literature on Bohr. This list is not complete and somewhat selective. This selection, however, is not primarily based on affinities between these works and this book, but on those works that offer comprehensive and engaging accounts of Bohr's thinking and work, sometimes from perspectives quite different from the one adopted here.

I thought that it would be most effective and helpful to the reader to follow the historical developments of quantum theory and Bohr's thinking, and I have arranged the chapters accordingly. Following Chap. 1, which serves an introduction, the subsequent chapters explore those junctures of Bohr's thinking at which his main concepts and arguments concerning quantum mechanics and complementarity emerged or were transformed, as manifest in his key published works, with the date of the publication chronologically marking each chapter. The first, introductory, chapter is an exception and is dated 1900–1962, the years, respectively, of Max Planck's discovery of quantum theory and Bohr's final interview on his life and work, literally on the eve of his death. (This interview itself will be addressed in Chap. 10). Chapter 2 is a kind of prologue to Bohr's thinking concerning quantum mechanics and complementarity, and briefly traces the development of this thinking to Bohr's 1913 theory of atomic constitution, which, as I shall argue, contains certain key ingredients of his later epistemological approach. The remainder of the book is devoted to Bohr's thinking following the discovery of quantum mechanics. It is not possible to completely map the development of Bohr's thought in this way, in part because some of the corresponding junctures of this development are discussed in later articles. For example, Bohr's initial discussions with Einstein in 1927, leading to major changes in his approach is only discussed by Bohr himself in his 1949 "Discussion with Einstein on Epistemological Problems in Atomic Physics," which presents the development of his thinking concerning quantum mechanics and complementarity via his lifelong debate with Einstein. Accordingly, the necessary adjustments and qualifications will be made in some of the chapters in order to accommodate these complexities. Nevertheless, I hope that the book's organization will help the reader to follow the trajectory of Bohr's thinking.

Acknowledgments

It would like to thank Giacomo Mauro D'Ariano, Christopher A. Fuchs, Gregg Jaeger, Andrei Khrennikov, Jan-Åke Larsson, N. David Mermin, and Theo Niewenhuizen, for productive discussions over the course of several years. My personal thanks to Marsha Plotnitsky, Rens Lipsius, and Inge-Vera Lipsius for their kindness and hospitality, and to Paula Geyh for her inspiring presence. I would like to thank to Martin Whitehead for his help with copy-editing the manuscript. I am grateful to Springer for publishing the book, and I would like to thank those with whom I worked at Springer, most especially Chris Coughlin for initiating and shepherding this project. A copy of the photograph of Bohr's drawing was kindly provided by the Emilio Segré Visual Archive of the American Institute of Physics and is reproduced here with their permission, which is gratefully acknowledged.

Contents

Abbreviation

PWNB   N. Bohr, *The Philosophical Writings of Niels Bohr*, 3 vols. (Ox Bow Press, Woodbridge, CT, 1987); N. Bohr, *The Philosophical Writings of Niels Bohr, Vol. 4: Causality and Complementarity, Supplementary Papers*, eds., J. Faye and H. J. Folse (Ox Bow Press, Woodbridge, CT, 1998)

Chapter 1
1900–1962: From Planck to Bohr

Quantum theory was initiated by Max Planck's discovery, in 1900, of his black body radiation law, which revealed, or in any event compelled Planck to postulate, that radiation, such as light, previously considered a continuous phenomenon in all circumstances, could also exhibit features of discontinuity or discreteness in certain circumstances. The limit at which this discontinuity appears is defined by the frequency of the radiation and a universal constant of a very small magnitude, h, Planck's constant, which Planck termed "the quantum of action" and which turned out to be one of the most fundamental constants of physics. The indivisible (energy) quantum of radiation in each case is the product of h and the frequency, $E = h\nu$. The role of h may be seen as analogous to the role of c (the constancy of the speed of light in a vacuum in its independence of the speed of the source) in special relativity, in terms of both the necessity of a departure from classical physics and of introducing the first principles of a new theory. Other earlier developments, sometimes referred to as the "old quantum theory," made apparent yet further complexities of quantum phenomena and posed new questions concerning them.

The old quantum theory was quite successful, sometimes spectacularly successful. It did have, however, major physical problems, which proved to be insurmountable for the theory and which quantum mechanics, discovered by Heisenberg in its matrix form and Schrödinger in its wave form in, respectively, 1925 and 1926, was able to solve. Quantum mechanics was further developed in

A. Plotnitsky, *Niels Bohr and Complementarity*, SpringerBriefs in Physics, DOI: 10.1007/978-1-4614-4517-3_1, © Arkady Plotnitsky 2013

the work of Born, Jordan, Dirac, Pauli, and (primarily in terms of interpretation) Bohr.[1] The theory was nonrelativistic and dealt with the motion of electrons at speeds significantly slower than those of light in a vacuum, although the initial work on relativistic quantum theory, quantum electrodynamics, was virtually contemporary, and Dirac introduced his relativistic equation for the electron in 1928.

While, however, quantum mechanics resolved most of the physical difficulties that beset the old quantum theory and brought with it a certain closure of nonrelativistic quantum theory by becoming the standard theory of the corresponding phenomena (which it has remained ever since), it also brought with it new, more radical epistemological complexities. To some, beginning with Einstein, these complexities were even more troubling than those of the old quantum theory. Bohr registered this fact by noting, in 1929, that in this regard quantum mechanics proved to be a "disappointment" to some, and it has remained a disappointment to many ever since.[2] He amplified this point later in "Discussions with Einstein" in speaking of "skepticism as to the necessity of going so far in renouncing customary demands as regards the explanation of natural phenomena" (Bohr 1949, PWNB 2, p. 63). The majority of even the most resistant critics, Einstein and Schrödinger among them, acknowledged the remarkable predictive capacity of the theory. What bothered these critics and even some proponents of quantum mechanics was the manifest deficiency of the explanatory-descriptive capacity of quantum mechanics as concerns the behavior of quantum objects. The situation was as follows.

---

[1] Most key founding papers on quantum mechanics are assembled in (van der Waerden 1968), and will be cited from this volume. Throughout this study, by "quantum mechanics" I mean the *standard* version of quantum mechanics, covered by Heisenberg's or Schrödinger's formalism, or other, more or less mathematically equivalent, versions of the formalism, such as those of Paul Dirac or John von Neumann. The term will not refer to alternative accounts of the experimental data in question, such as those offered by Bohmian mechanics, for example (and there are several versions of the latter). "Quantum theory" will refer more generally to theoretical thinking concerning quantum phenomena. This denomination, thus, includes the old quantum theory; alternatives to standard quantum mechanics, such as those just mentioned; and higher level quantum theories, such as quantum electrodynamics and other quantum field theories. "Quantum phenomena" will refer to those observable phenomena in considering which the role of Planck's constant h cannot be neglected, as it can be in the case of the phenomena considered by classical physics. "Quantum physics" will refer to the overall assembly of experimental and theoretical accounts of these phenomena. "Classical mechanics" (or "Newtonian mechanics"), "classical theory," and "classical physics" will be used along respectively parallel lines.

[2] "The Quantum of Action and the Description of Nature," in Philosophical Writings of Niels Bohr, 3 vols (Bohr 1987, vol. 1, p. 92). This collection will hereafter be referred to as PWNB, followed by a volume number. A supplementary volume of Bohr's essays was published as Niels Bohr, *The Philosophical Writings of Niels Bohr, Volume IV: Causality and Complementarity, Supplementary Papers*, J. Faye and H. Folse, ed. (Bohr 1998), and it will be referred as PWNB 4. These four volumes contain most of Bohr's works on quantum mechanics and complementarity to be cited here. The key articles are listed separately in the bibliography with the original publication date, to be given in the text as well: thus, the article just cited will be referred to as "Bohr 1929a, PWNB 1," followed by page numbers.

The old quantum theory dealt reasonably well with statistical matters and was or, rather seemed, to be analogous to classical statistical physics. What was lacking was the mechanics describing the behavior of individual quantum objects and that would, thus, underlie the manifest statistical behavior of multiplicities in a manner similar to Newtonian mechanics, which explains causally the motion of individual molecules in classical statistical theory. Bohr's 1913 theory of the hydrogen atom and its subsequent extensions, Einstein's 1916 work on so-called induced and spontaneous emissions, and several other earlier theories did deal with individual quantum processes, which were encountered already in the case of radioactivity.[3] However, while describing well the atomic spectra in terms of the discontinuous transition ("quantum jumps") of electrons in the atom from one energy level to another, Bohr's theory did not account for the mechanism of this transition. Indeed, the theory was incompatible with classical mechanics and classical electrodynamics alike. Einstein's arguments just mentioned suggested that probability might be irreducible even in considering individual quantum processes (rather than those involving large multiplicities of quantum objects), yet one more of Einstein's profound insights and a defining feature of quantum theory, which, however, was also to contribute to Einstein's discontent with quantum mechanics. This point was amplified by the 1924 proposal of Bohr, Hendrik Kramers, and John Slater (the BKS proposal), which aimed to resolve some of the difficulties of the old quantum theory by suspending the strict application of the energy conservation law in quantum processes, where the law would only apply statistically (Bohr et al. 1924). The proposal, controversial to begin with, was abandoned in view of the experimental findings, especially those by Walther Bothe and Hans Geiger, which confirmed the exact energy conservation in quantum interactions. These works, thus, further showed that true mechanics, accounting for individual quantum processes, was lacking.

The new quantum mechanics, which was expected to resolve these problems was, however, nothing like classical mechanics, and was not a theory that was expected or hoped for by most at the time either. The new quantum theory could only predict, in general probabilistically, the outcome of certain events, such as collisions between a particle and a silver-bromide photographic screen, but it appeared unable to describe the motion of quantum objects in a manner analogous to classical mechanics. Indeed, the latter predicts because it describes, causally, the behavior of classical objects, the situation found in relativity as well. In classical statistical physics, too, the underlying individual processes are treated as causal, even though the theory does not describe them as such, which, as just noted, could not be properly maintained even in the old quantum theory, beginning with Planck's law. It became clear almost immediately in the wake of Planck's discovery of his law, however, that this law was incompatible with a classical-like underlying picture. For that reason, the way of statistical counting is different in classical and quantum statistics. Planck's counting was correct, even though part of his physics was wrong, an error that, as

---

[3] On Einstein's work on this subject, see (Pais 1982, pp. 402–14) and (Pais 1991, pp. 191–192).

Einstein observed, was most fortunate for physics (Einstein 1949a, pp. 37–43). Quantum mechanics provided little, if any, help in making it possible to assume causality at the level of elemental individual quantum behavior, such as that of individual electrons and photons. If anything, it suggested that the indeterministic character of our quantum predictions may be due to the lack of such an underlying causality of the individual quantum processes, rather than our practical difficulties, found also in classical physics, of properly tracking the corresponding causal dynamics. (I shall define "causality" and "determinism," and explain the difference between them in detail below.)

Quantum mechanics did not eliminate the possibility that one could establish a causal or, in the first place, realist (descriptive) model of the elemental quantum behavior. The hope, the Einsteinian hope, that such a theory could eventually be discovered has continued to persist, and still does, sometimes linked to hopes or even claims for causal interpretations of quantum mechanics itself. That nature might not (not the same as it will not!) allow for such a theory is a minority view, which was, however, taken by Bohr and defined his debate with Einstein. Peculiarly, under certain conditions, collective quantum behavior does exhibit a certain correlational order not found in classical physics. Thus, quantum mechanics does not proceed in the way classical statistical physics does, from causal individual behavior to the collective behavior that only allows for probabilistic predictions. Instead, it combines or rather responds to a combination, found in nature, of the irreducibly lawless individual behavior with a statistically correlated, ordered collective behavior. This combination is one of the great enigmas of quantum physics, an enigma that may not allow a solution on classical lines because it may not be possible to conceive of how such a combination may come about.

The situation just outlined was, as far as most doubters, beginning, again, with Einstein, were concerned, not helped by Bohr's attempt to interpret this situation via his concept of complementarity, introduced in 1927. This interpretation was welcome by some close to Bohr, most especially Heisenberg and Pauli, and accepted as logically consistent by most, including Einstein, although, while he generally grasped Bohr's epistemology, Einstein, as I noted, was ill at ease with the concept of complementarity itself, the "sharp meaning" of which, he said, "eluded" him (Einstein 1949b, p. 674). However, Bohr's interpretation, at least in the versions developed, under the impact of his discussions with Einstein, after the Como lecture, and especially in its ultimate version (in place by the late 1930s) offered little to alleviate epistemological discontents concerning quantum mechanics. This was because, rather than removing or at least mitigating the epistemological difficulties indicated above, this interpretation based its logic on these difficulties, seen by Bohr not as problems but as solutions.

First of all, Bohr's interpretation was based on the impossibility of having a proper (by the standards of classical physics or relativity) hold on the actual behavior of individual quantum systems and to offer a realist description of this behavior. Second, in effect as a consequence, the irreducibly noncausal character of this interpretation made randomness and the recourse to probability unavoidable even in the case of primitive (indecomposable) individual processes and events, as opposed to classical mechanics, which is, correlatively, both a realist and causal theory of individual classical processes, *at least in principle and in the case of idealized models*. The underlined qualification is important. From Galileo and Newton on, modern physics deals only with such idealized models, which then can be used in considering actual physical objects. Such models appear to be difficult, and in Bohr's view, especially his ultimate view, are impossible, to develop in quantum physics. With the help of his concept of complementarity, Bohr proceeded to build his interpretation of quantum mechanics on both features—the lack of realism and the lack of causality in the quantum–mechanical description of quantum objects and processes—again, seeing them not as problems but as solutions.

This was not a solution that was palatable to Einstein and others who shared his concerns. In addition, Einstein and others argued (at the earlier stages of the debate, correctly) that certain questions concerning the quantum–mechanical account of these phenomena remained unresolved. It took Bohr another decade and, in some respects, even longer to resolve some of these difficulties, although not many, beginning, again, with Einstein, accepted this resolution as philosophically satisfactory. The situation led to an intense debate, in particular to the great confrontation between Einstein and Bohr. This confrontation began in 1927, and it has overshadowed the history of quantum mechanics and the continuing debate concerning it ever since. Einstein never accepted the view advocated by Bohr as corresponding to the ultimate constitution of nature, and his rejection has been a source of comfort and inspiration for many others. Most famous is Einstein's refusal to believe that God would resort to playing dice or rather to playing with nature in the way quantum mechanics appeared to suggest, which is indeed quite different from playing dice. According to his later (around 1953) remark, a lesser known or commented upon, but arguably more important one: "That the Lord should play with dice, all right; but that He should gamble according to definite rules, that is beyond me" (cited in Wheeler and Zurek 1983, p. 8). Einstein clearly refers to the rules of quantum mechanics as essentially different from the rules for calculating probabilities for regular dice throwing. He was unwilling to accept that quantum mechanics ultimately reflects nature, and he thought that an alternative theory that would be more philosophically palatable and, by his criteria, more complete could eventually be found. He thought that quantum mechanics revealed a beautiful element of truth in nature, but not its ultimate truth, and he thought that "it offers no useful point of departure for future development" [of theoretical physics] (Einstein 1949a, p. 83).

Einstein's or others' preference for more classical-like alternatives to quantum mechanics is perfectly legitimate. It is certainly not my aim, nor was it Bohr's, to deny that this view is possible, and I do not claim that Bohr offered more than a particular interpretation of quantum phenomena and quantum mechanics, one among many possible interpretations. The question is whether nature allows for more classical-like alternatives, desired by Einstein and many others, and Bohr's view was that it just might not. This question remains open, since we have not heard the last word from nature on this, that is to say, the next word, the only last word possible in physics. Quantum mechanics has been around for nearly a century now, and within its proper scope, has remained the standard theory, well-confirmed experimentally, as are other currently standard quantum theories, in particular, quantum electrodynamics, arguably the best confirmed physical theory ever, and other quantum field theories. Besides, that physics will move on in the direction of an even greater difference from Einstein's desiderata is not inconceivable, either.

I would like now to explain Bohr's concept of complementarity. First, however, I shall explain what I mean (in part following G. Deleuze) by "concept," a term often used without further explanation in physics and even in philosophy. A concept is a multi-component entity, defined by the specific organization of its components, which may be (more or less) general or (more or less) particular in nature. It is the relational organization of a concept's components, the architecture of a concept, that is most crucial, and if, within the same organization, one replaces certain more general elements with their particular instances, one could speak of an instantiation of this concept, although such an instantiation is still a concept in its own right. It is rare for a concept to have only one component. Consider the concept of "tree," even as it is used in our daily life. One the one hand, it is a single generalization. On the other hand, what makes it that of "tree" is the at least implied presence of further elements, components, or subconcepts, such as "branches," "roots," and so forth. The concept of tree in botany acquires further features and components, indeed becomes a different concept, and it becomes a still different concept in biology. The concept of a moving body in classical physics is manifestly multi-component, since it involves multiple (idealized) elements related to its properties and behavior, such as and in particular, its motion. This concept is very different from that of a quantum object, the properties of which cannot, in Bohr's view, be specified in the way they are in classical physics. A single-component concept is, thus, usually the product of a provisional cutoff of its multi-component organization.

Although the concept of complementarity was not quite given by Bohr the exact definition about to be formulated, this definition may be surmised from several of his descriptions of the concept. In this regard, Bohr was not always helpful to his audience, beginning, again, with Einstein who, as noted above, thought that the sharper meaning of complementarity eluded him, after decades of debating quantum mechanics with Bohr. I do think, however, that Bohr's reply to the argument of Einstein, Podolsky, and Rosen (the EPR argument), a reply that, as will be seen in Chap. 8, Einstein misread, gives complementarity this type of meaning, if, again, not in a single sharp formulation. Bohr comes closest to such a single formulation in "Discussion with Einstein": "The evidence obtained under different experimental conditions cannot be comprehended within a single picture, but must be regarded as complementary in the sense that only the totality of the phenomena exhaust the possible information about the [quantum] objects" (Bohr 1949, PWNB 2, p. 40). Complementarity, then, is defined by

(a) a mutual exclusivity of certain phenomena, entities, or conceptions; and yet
(b) the possibility of applying each one of them separately at any given point; and

(c) the necessity of using all of them at different moments for a comprehensive account of the totality of phenomena that we must consider.

This definition is very general and allows for different instantiations of the concept in the case of quantum phenomena and for the application of the concept beyond physics, as discussed in Chap. 9. (We keep in mind that these instantiations are concepts in their own right.) Part (b) is not stated in the above formulation from "Discussion with Einstein" either. It can, however, easily be established on the basis of Bohr's other elaborations there, such as the one, via the Compton effect, immediately following Bohr's formulation cited above (Bohr 1949, PWNB 2, p. 40) or elsewhere, especially in his reply to EPR, where, as will be seen, Bohr's argument essentially depends on this possibility. Parts (b) and (c) of this definition are just as important as part (a), and to miss them, as is often done, is to miss much of the import of Bohr's concept.

As I shall discuss in Chap. 4, Bohr "helped" this misunderstanding by his Como argument, which introduced the concept. The instantiation of complementarity as that of the space–time description, obtained by observation, and the causality of independent, "undisturbed," quantum behavior, central to the Como argument (but gone from Bohr's writings shortly thereafter), does not properly conform to this definition. Arguably the most significant instantiations of complementarity rigorously following this definition are those of the complementary nature of position and momentum measurements, and of the space–time coordination and the application of momentum or energy conservation laws (there are, thus, two complementarities here), correlative to Heisenberg's uncertainty relations (e.g., Bohr 1958, PWNB 3, p. 5). These instantiations became central to Bohr's thinking, following his initial exchanges with Einstein in 1927 and extending to the rest of his work on quantum theory.

Technically, as expressed by the famous formula, $\Delta q \, \Delta p \cong h$ (where q is the coordinate, p is the momentum, and Δ is the root-mean-square deviation of the value of a given variable from its mean value and h is Planck's constant), the uncertainty relations may only appear to prohibit the simultaneous exact measurement of both the coordinate and the momentum associated with a given quantum object.[4] Importantly, the uncertainty relations apply regardless of the capacity of our measuring instruments: that is, the uncertainty relations would be valid even if we had ideal instruments. In other words, in practice, in classical and quantum physics alike, one can only measure or predict each variable with arbitrary precision within the capacity of our measuring instruments, and in classical physics one can, in principle, measure both variables simultaneously. The uncertainty relations would prohibit us from doing so for both variables regardless of this capacity. This fact is well confirmed experimentally, thus making the uncertainty relations a law of nature. In Bohr's interpretation, however, the uncertainty

---

[4] An uncertainty relation applies to the coordinate and the momentum in a given direction. For a quantum object in three-dimensional space, each quantity will have three components, defined by the chosen coordinate system.

relations imply something more. They make each type of measurement mutually exclusive to each other. These measurements are complementary in Bohr's sense. They can never be performed simultaneously, and yet it is always possible to perform either type of measurement separately at any given point, and it is necessary to use both types of measurements at different moments for a comprehensive account of the totality of phenomena, quantum phenomena, that quantum mechanics must consider. Indeed, in Bohr's interpretation of the uncertainty relations, one not only cannot measure both variables simultaneously but also can never define both simultaneously for the same quantum object. The joint simultaneous measurement and definition of both variables is, however, always possible, at least in principle in the case of idealized models, in classical physics, and it is this possibility that allows one to maintain causality there. Accordingly, in Bohr's view, the uncertainty relations reflect the difficulty and ultimately impossibility of making a theory accounting for quantum phenomena causal or, again, realist, to begin with, and in Bohr's interpretation quantum mechanics is neither.

Although Heisenberg's introduction of the uncertainty relations was a major impetus to Bohr's invention of complementarity, it does not appear to be the only one, and, unlike in the late instantiations of the concept just mentioned, the uncertainty relations had a more indirect connection to complementarity in the Como argument. Bohr's earlier approaches to complementarity (he made some moves in this direction before the uncertainty relations were introduced) were shaped by the apparently necessary use of certain mutually exclusive conceptions in quantum theory, such as those of particles and of waves, and the corresponding physical theories, within the same theoretical framework. This is in part why Schrödinger's wave mechanics as a whole, rather than only his equation, initially had a certain, qualified, appeal to Bohr. Even at this stage of his thinking, however, Bohr was aware of the difficulties of applying the idea of waves to quantum objects themselves, even by way of complementarity; he avoids the language of waves and, more crucially, the concept of wave-particle complementarity altogether. I would argue that wave-particle complementarity, with which the concept of complementarity is arguably associated most, did not play a significant, if any, role in Bohr's thinking, at least after the Como lecture and even there, Bohr's ultimate solution to the dilemma of whether quantum objects are particles or waves—or his "escape" from the paradoxical necessity of seeing them as both—is that they are neither. This view is perfectly legitimate and consistent, once either type of feature or, again, either type of effect (particle-like or wave-like), is transferred to the level of measuring instruments and once these effects are recognized as mutually exclusive.

Also, influenced by Schrödinger's approach and the Dirac–Jordan transformation theory (which brought both versions of quantum mechanics together), Bohr, as noted above, briefly entertained the idea that the independent behavior of quantum objects is causal and that the lack of causality in the observable quantum phenomena was due to the "disturbance" of this behavior unavoidably introduced by measurement. He proposed the idea in 1927 in the Como lecture, to be discussed here in Chap. 5. Indeed, he used this idea to define the concept of complementarity, instantiated by the mutually exclusive nature of the space–time coordination (provided by an act of observation, which uncontrollably "disturbs" the behavior of quantum objects) and the claim of the causality of independent, "undisturbed," behavior of quantum objects. Bohr, however, quickly abandoned the idea, in part under the impact of his initial exchanges with Einstein in 1927. Bohr was compelled to turn to different instantiations of the concept, correlatively to a more radical epistemology of quantum phenomena and quantum mechanics, an epistemology that also places an additional emphasis on the role of randomness and chance, and hence probability in quantum theory. I would now like to briefly outline the relationships among ontology (or reality), epistemology, causality (or the lack thereof), and probability in Bohr's thinking. First, I shall define these terms themselves, as they will be understood here (they can be understood otherwise).

As I said, in Bohr's view quantum objects are seen as indescribable and unknowable, and in his ultimate view even inconceivable, or as idealizing the ultimate constitution of nature accordingly. They only manifest their existence in certain effects of their interactions with measuring instruments upon those instruments. Nevertheless, Bohr assumes the independent existence of quantum objects or, again, something in nature idealized in this way. It is the existence of such entities (also as something that had existed before we appeared in the Universe and that will exist when we will no longer exist) that is responsible for quantum phenomena, which are defined by the effects of the interactions between these entities and measuring instruments, effects available to our perception, knowledge, and thought. Although held by Bohr for general philosophical reasons, this assumption becomes especially important in Bohr's exchanges with Einstein concerning EPR-type experiments and the question of locality (the absence of instantaneous physical action at a distance, incompatible with relativity) these experiments pose. Quantum objects may be said to be real, although it may not be possible to have a realist theory of them, even by way of idealization, and it would, again, be difficult to have any theory other than one defined by idealized models, even in classical physics.

By realist theories I understand theories of the following two types. This understanding appears to be sufficiently comprehensive to absorb most uses of the term, at least those to be discussed in this study. According to the first type of realism, a realist theory would offer an actual mapping of the properties of the physical systems considered and their behavior, which mapping may be exact or approximate, or displaced, as in Bohmian mechanics, for example. According to the second type of realism, one would presuppose an independent architecture of reality governing this behavior, even if this architecture cannot be mapped by a theory, however partially or approximately, at a given point of history and perhaps even ever, but if so, only due to practical limitations. In this case, a theory that is merely predictive may be accepted for lack of a realist alternative, but usually under the assumption that, enabled by our conceptual and mathematical imagination, a future theory will do better. This view defined, for example, Einstein's attitude towards quantum phenomena or objects. What unites both conceptions of realism and ultimately defines realism most generally is the assumption, along Kantian lines, that this type of architecture, which may be temporal, objectively exists. In other words, realism is defined by an assumption that the ultimate constitution of nature possesses attributes that may be unknown or even unknowable, but that are thinkable, conceivable (Kant 1997, p. 115). In particular, it is deemed conceivable in the model of classical physics and its idea of physical reality, assumed to be at least partially approachable in an idealized way through the mediation of the mathematized concepts of classical physics. At the very least, realism assumes that the concept of organization can in principle apply to this constitution, no matter how much off the mark anything we can specifically come up with in conceiving of this constitution may be. The hope, however, is that we can capture, even if, again, approximately and by way of idealized models, something of this architecture. This was more or less Einstein's position, which is why it is often insisted that he was not a naïve realist.

Bohr's epistemology is not based on any of these assumptions, in particular on the assumption that a description of or even conception of quantum objects and their independent behavior is possible. Moreover, at least in Bohr's ultimate version of his epistemology, these assumptions are expressly disallowed. Bohr does, again, assume that quantum objects or some entities in nature that they thus idealize exist independently of our interaction with them, and that it is their existence that is responsible, through this interaction, for this situation. In other words, the character of the existence of such entities is such that it disallows our describing, or even forming a conception of this nature, at least, again, in Bohr's ultimate view of the situation. There are intermediated stages of Bohr's interpretation when certain partial conceptions of quantum objects are allowed. The lack of causality is an automatic consequence, given that quantum objects and processes are beyond any description and even conception. Causality would be a feature of such a description and hence is disallowed automatically, as Schrödinger noted in his cat paradox paper, by way of a very different assessment of this type of argumentation, which he saw as "a doctrine born of distress" (Schrödinger 1935, pp. 152, 154).

I understand "causality" as an ontological category (part of reality) relating to the behavior of physical systems whose evolution is defined by the fact that the state of a given system is, at least, again, at the level of idealized models, exactly determined at all points by their state at a particular point, indeed at any given point.[5] I understand "determinism" as an epistemological category (part of our knowledge of reality) that denotes our ability to predict the state of a system, at least as defined by an idealized model, exactly (rather than probabilistically) at any and all points once we know its state at a given point. Again, usually the knowledge of the state at any point suffices. Determinism is sometimes used in the same sense as causality is used here. However, while it follows automatically that noncausal behavior, again, considered at the level of a given model, cannot be

---

[5] It is usually assumed that a cause precedes its effect, or is simultaneous with it. Relativity further restricts the application of the concept of causality by the assumption that causal influences cannot travel faster than the speed of light in a vacuum, c, the assumption known as the principle of locality. Sometimes, the term "causality" is used in this sense. When speaking of the lack of causality in quantum mechanics, I only mean the inapplicability of the concept of causality found in classical physics and not any form of incompatibility with relativity.

handled deterministically, the reverse is not true. The underlined qualification is necessary because we can have causal models of processes in nature that may not be causal. Thus, while the causal models of classical physics apply and are effective within the proper limits of classical physics, this does not mean that the ultimate character of the actual processes that are responsible for classical phenomena are causal. For example, an electron sufficiently far away from the nucleus of an atom can sometimes be treated as a classical object moving along an orbit, which, however, does not mean that this is how it actually behaves, since the situation is more consistently treated as quantum and, hence, likely noncausal.

It is true that noncausal models of the quantum constitution of nature are also idealizations, which may be necessary (not everyone agrees) given the nature of quantum phenomena in the corresponding interpretation. Rigorously, it is only determinism that these phenomena prevent, by virtue of the fact, well-established experimentally, that identically prepared quantum experiments in general lead to different outcomes. In other words, the causal or, conversely, noncausal character of our models does not guarantee that the actual behavior of quantum objects is causal or noncausal, and in this sense causality may be seen as an epistemological concept as well. It is, however, useful to retain both concepts, causality and determinism, and the difference between them at the level of models in order to capture the essential difference, at least according to Bohr's view, between classical and quantum physics. In classical physics, our predictions, deterministic or probabilistic, are made on the basis of realist (descriptive) causal models of the behavior of the individual processes involved. In quantum physics, an introduction of causal or, and to begin with, realist models even for primitive (undecomposable) individual quantum processes at the very least poses major difficulties and may not be possible.

Thus, classical mechanics (excluding the chaos-theoretical part of it) deals deterministically with causal systems, at least, again, at the level of idealized models, since there are practical limitations upon the measurements involved. Classical statistical physics deals with causal systems, but only statistically, in terms of probabilities, rather than deterministically. Chaos theory deals with systems that are, in principle, causal, but whose behavior cannot be predicted even in statistical terms in view of their nonlinearity and, hence, their sensitivity to initial conditions. These and other classical physical theories (those of electromagnetic radiation and other continuous phenomena included) are causal insofar as they deal, deterministically or not, with (idealized) systems that are assumed to behave causally. As explained above, these theories, at least in most versions or interpretations, are also, and in part correlatively, realist on the definition given above, and as Ludwig Wittgenstein noted, conversely, any actual realist description may have to be causal (Wittgenstein 1924, p. 175). Both classical mechanics (exclusive of its chaos-theoretical part) and chaos theory are realist insofar as such a mapping is assumed to take place, at least as an idealization or as a good approximation that such models provide, although chaos theory is not deterministic. By contrast, classical statistical physics is not realist in the same sense insofar as its equations do not describe the behavior of its ultimate objects, such as molecules of a gas. It is, however, generally based on the realist assumption of an underlying nonstatistical multiplicity, whose individual members in principle conform to the causal laws of classical, Newtonian mechanics of individual physical systems. Although this assumption involves certain conceptual difficulties, it is generally maintained in classical statistical physics. It may be added that, while relativity is a causal theory of the individual system considered as well, it poses problems as concerns a realist description of such system, in particular photons (ultimately a quantum objects, however), on which I shall comment later in this study. Initially, it appeared possible to adopt the assumption that individual quantum objects behave causally in quantum theory as well, even if with certain qualifications, apparently necessary in view of Planck's discreteness postulate and Bohr's postulates, in part based on Planck's postulate, concerning atomic structure.

By "randomness" or (they are not quite the same) "chance" I refer to a manifestation of the unpredictable. A random or chance event is an unpredictable event. It may or may not be possible to estimate whether such an event would occur, or often to anticipate it. Physically, such an event may, as in classical statistical physics, or may not, as in quantum mechanics (at least in Bohr's interpretation), hide some ultimate causal dynamics. In the latter case, chance becomes irreducible in principle rather than only in practice, as in the case of classical statistical physics, which makes this theory merely indeterministic rather than noncausal.

Probability and statistics deal, theoretically or practically, with providing estimates, possibly numerical, of occurrences of certain individual or collective events, in quantum mechanics or in physics, or science in general in accordance with mathematical probability theories, which theorize such estimations. Although related and often used interchangeably, the terms probabilistic and statistical are generally different. Probabilistic refers to our estimates of the probabilities of either individual or collective events, such as of a coin toss or of finding a quantum object, such as an electron or a photon, in a given region of space. Statistical refers to probabilistically estimating the behavior of identical objects in a large number of repeated trials or of systems composed of a large number of individual constituents. Thus, the term "quantum statistics" refers to the behavior of large multiplicities of identical quantum objects (e.g., quantum gas), such as electrons and photons, which behave differently in such multiplicities than do classical objects. On the other hand, in Bohr's interpretation, quantum mechanics is a probabilistic theory of individual quantum events. While chance introduces the element of chaos into order and reveals the character of the world, or of our interactions with the world, as the interplay of order and chaos, probability introduces an element of order into situations defined by the role of chance, and allows us to handle such situations better. Probability is, thus, as much about order, about patterns, as it is about randomness, or about the interplay of randomness and patterns, which aspects of probability acquire a special significance in quantum physics, in view of the presence of correlations and hence a form of order between certain quantum events.

Attempts to reintroduce causality in quantum theory have never died and they still continue. This is not surprising given the history of causality from the pre-

Socratics on and especially the dominance of the causal ontology of just about every process considered in modern science, philosophy, and most other fields, especially in the eighteenth-century Enlightenment, sometimes also known as the age of reason. Most of the key philosophical views that dominate our culture were established or crystallized then. A causal ontology of the world was assumed by Kant, who was inspired by Newton's physics and who was largely responsible for the term "Enlightenment." It appears (there is some debate concerning this) that this type of ontology was also assumed by Hume, who was generally more skeptical concerning the validity of the concept of causality. What they denied was that the human mind could have a full access to this causality and thus establish definitive causal connections between events, rather than only surmise probable connections between events. As Bohr noted, in physics and philosophy alike "the unrestricted applicability of the causal mode of description to physical phenomena has hardly been seriously questioned until Planck's discovery of [quantum phenomena]" (Bohr 1938, PWNB 4, p. 94; on philosophy, Bohr 1949, PWNB 2, p. 65). One could think of a few exceptions in philosophy, such as Friedrich Nietzsche. Einstein was, again, the most prominent advocate of the view of nature as ultimately casual in the twentieth century, which he expressed, as part of his criticism of quantum mechanics, by his famous pronouncement "God does not play dice" (e.g., Letter to Born, December 4, 1926, Born 2005, p. 88). He never wavered in his belief in a classical-like approach to physical reality, although he held a complex position on how an access to this reality is possible. He thought, close to Kant and Hegel, that it is only possible through a free choice and, in the first place, creation of concepts, rather than on the basis of observations alone, along empiricist or positivist lines, such as those advocated by Ernst Mach.[6]

In any event, at least in Bohr's interpretation, quantum mechanics is neither, more manifestly, a deterministic nor, more crucially, a causal or, in the first place, realist theory. It is not causal, first, insofar as a given quantum event or

---

[6] The terms "empiricism" and "positivism," and the philosophical views with which they are associated have a long history and encompass a broad spectrum of positions. They do not, in general, have the same meaning, although their meanings are related and sometimes converge, as in the case of logical positivism, also known as logical empiricism. Roughly, empiricism, especially as initiated by the philosophy of John Locke, is a theory of knowledge based on the claim that knowledge derives most essentially from sensory experience. Positivism, especially as a form of philosophy of science, commonly adds logical and mathematical treatment of the empirical data as the second essential source of all real knowledge. See, for example, the corresponding entries in Stanford Encyclopedia of Philosophy (http://plato.stanford.edu/) for further discussion. Whatever the further complexities of empiricism and positivism, or of Mach's philosophy, the most essential point of difference between Einstein's realist thinking or, in this respect, Bohr's nonrealistic thinking and positivist or empiricist views is defined by the fundamental role, for both Einstein and Bohr, of concepts and hence of human thought in scientific knowledge. For Einstein, the creation or use of concepts always stands in the service of scientific realism, but it need not do so, and it does not for Bohr. Einstein's view is close to Kant and his followers, especially Hegel, who in part advanced their views of the role of concepts in theoretical thinking against empiricism, such as that of Locke and Hume. For Kant and Hegel, thought and concepts shape and even determine what is, or appears to be, observed by senses.

phenomenon, manifest, ontologically and phenomenally, in our measuring instruments, cannot be determinately and especially continuously physically connected to other such events. At least, such a connection can never be guaranteed at the level of individual events, and can never be established by describing or even conceiving of a continuous physical process responsible for this connection. This view is correlative to, and in fact was Bohr's epistemological radical interpretation of, the circumstance that the identically prepared quantum experiments, as concerned the manifest, classical state of our measuring instruments, in general lead to different outcomes, which is, as noted above, arguably the defining manifest feature of individual quantum phenomena. This identical preparation is always possible, as possible as it is in classical physics; indeed it is possible because the manifest physical state of measuring instruments can be treated classically. Accordingly, a theory properly accounting for these phenomena cannot be deterministic, although, in contrast to quantum mechanics in Bohr's interpretation, such as theory may in principle be causal and causally realist as concerns the independent behavior of quantum objects that connects such observed phenomena. For, this circumstance itself does not exclude the possibility that the underlying physical processes responsible for quantum phenomena are ultimately causal. Bohmian theories, which are mathematically different from quantum mechanics but which reproduce exactly the probabilistic predictions of quantum mechanics (and are also equally consistent with the uncertainty relations), are such theories. These theories are, however, nonlocal: they imply instantaneous physical action at a distance, which is incompatible with relativity. It is sometimes also claimed by the proponents of standard quantum mechanics that independent quantum processes are causal and even that quantum mechanics in fact causally describes them. In this view, the lack of determinism in our predictions concerning observable quantum phenomena is due to the role of measurement, which "disturbs" the causal character of independent quantum processes. This view is far from uncommon, and as indicated above and as will be discussed in Chap. 4, Bohr briefly, but only briefly, subscribed to this view, and based in it his first example of complementarity, as that of the space–time coordination, defined by measurement, and the claim of causality, possible if quantum processes are considered independently.

It may be added that, in Einstein's view, even under the assumption of the causal nature of independent quantum processes, insofar as quantum mechanics only predicts the probabilistic outcomes of experiments, it would only be a correct but not a complete theory, because a proper description of and exact predictions concerning individual quantum processes would still be lacking. At least, as will be seen in Chap. 8, Einstein contended that such is the case if one assumes locality, which Einstein saw as imperative. In part for that reason, Bohmian mechanics was unacceptable to him, in spite of its realist and causal character. Bohr, in responding to Einstein's contention, argued that quantum mechanics could be seen as a complete theory, as complete a theory of quantum phenomena as possible, without giving up locality. What has to be given up instead is causality and, in the first place, realism, another key imperative of Einstein. For, it does not appear possible and, in Bohr's view, is rigorously impossible to apply the concept of state as conceived in classical physics to quantum objects and processes, which makes the suspension of causality automatic. It follows that even in the case of primitive individual quantum processes and events we are left only with probabilistic predictions concerning them, in absence, moreover, of any description or even conception of their character. Nature, Bohr contended (Einstein was not convinced), makes quantum randomness and chance irreducible even in this case, at least, again, insofar as the data in question remain in place.

As noted earlier, under certain circumstances found in quantum experiments, such as the double-slit experiment or experiments of the EPR type, we encounter correlational and hence partially ordered patterns of collective quantum phenomena or events. These patterns are fundamentally quantum (they are not found in classical physics), and the rules for predicting them are encoded in the formalism of quantum mechanics. This is one of the reasons for Einstein's puzzlement, cited above, concerning the possibility that the Lord, if he plays dice, "should gamble according to definite rules [of quantum mechanics], that is beyond me" (Wheeler and Zurek 1983, p. 8). It was, however, not beyond Bohr, whose thinking concerning quantum phenomena and quantum mechanics saw the presence of this type of gamble in nature or our interactions with nature, as the solution of the quantum puzzle, as Einstein sometimes called it, as well. Or rather, it was beyond Bohr's thinking, too, and in Bohr's view, it is, by definition, beyond everyone's thinking, insofar as how this is possible is beyond the reach of our thought. However, unlike Einstein, who, for this reason, questioned quantum mechanics, at least as establishing a proper ground for a future development of physics, Bohr was ready to accept this "beyond" as a solution of the quantum puzzle, a solution that tells us that this puzzle cannot be solved by finding out how the emergence of quantum phenomena is possible. Bohr thought this "beyond-thought" to be a necessary counterpart of what we can and must think and know in quantum physics, and as a way of solving many problems and puzzles of quantum theory, which could not have been solved otherwise. This view carried with it a radical transformation of our understanding of nature, and, I would argue, of the character of scientific knowledge and, ultimately, of all human knowledge.

This transformation takes a lot of thought, a lifetime of thought. In responding, in his final interview conducted in 1962, to Thomas Kuhn's question "How did problems of this sort come to you in the first place?" about a very early (around 1903) anticipation, via Riemann's theory of the functions of complex variables, of the concept of complementarity, Bohr said: "I don't know. It was in some way my life, you see" (Bohr 1962, Session 5). That was never to change, and was made more intense when Bohr's thinking moved to quantum theory. It was a life of thought.

Chapter 2
1913. "On the Constitution of Atoms and Molecules": Quantum Jumps and Epistemological Leaps

In the spring of 1913, Niels Bohr completed his, now famous, paper, "On the Constitution of Atoms and Molecules," presenting a new model of the hydrogen atom. The paper built on previous discoveries of Planck and Einstein, mentioned in Chap. 1, concerning the quantum behavior of light. Along with these discoveries, Bohr's paper revolutionized our understanding of the ultimate nature of both radiation, such as light, and matter, such as electrons. Bohr, back in Denmark after a few years in England, sent the manuscript to his former mentor, a great physicist in his own right (a Nobel Prize laureate by then) Ernest Rutherford, a New-Zealander by birth, but long a member of the British scientific establishment. Rutherford was also the editor of *Philosophical Magazine*, a leading *physics* journal, founded a century earlier, where Bohr wanted to publish the paper and its sequels already in preparation. I shall return to the irony of the journal's title, which was unlikely to escape Bohr, below.

Upon reading the paper, Rutherford, first of all, made an important substantive comment reaching to the core of Bohr's argument, a comment that I shall discuss below. He also added, however, a "criticism of minor character" concerning "the arrangement of the paper":

> I think in your endeavour to be clear you have a tendency to make your papers much too long, and a tendency to repeat your statements in different parts of the paper. I think that your paper really ought to be cut down, and I think this could be done without sacrificing anything to clearness. I do not know if you appreciate the fact that long papers have a way of frightening readers, who feel that they have not time to dip into them. ... I will go over your paper very carefully and let you know what I think about the details. I shall be quite pleased to send it to *Phil. Mag.* but I would be happier if its volume could be cut down to a fair amount. In any case I will make any corrections in English that are necessary. ... I shall be very pleased to see your later papers, but please take to heart my advice, and try to make them as brief as possible consistent with clearness. ... P.S. I suppose you have no objection to my using my judgment to cut out any matter I may consider unnecessary in your paper? Please reply. (Letter to Bohr, March 20, 1913, reproduced in "The Rutherford Memorial Lecture," PWNB 3, p. 41)

A. Plotnitsky, *Niels Bohr and Complementarity*, SpringerBriefs in Physics, DOI: 10.1007/978-1-4614-4517-3_2, © Arkady Plotnitsky 2013

In commenting on this criticism in "The Rutherford Memorial Lecture" in 1958, Bohr said: "[This] point raised with such emphasis in Rutherford's letter brought me into a quite embarrassing situation. In fact, a few days before receiving his [letter] I had sent Rutherford a considerably extended version of the earlier manuscript. ..." (*PWNB* 3, p. 42). Rutherford tried to reason with Bohr again in responding to an expanded version of the paper: "The additions are excellent and reasonable, but the paper is too long. Some of the discussions should be abbreviated. As you know it is the custom in England to put things very shortly and tersely, in contrast to the German method, where it appears to be a virtue to be as long-winded as possible" (Letter to Bohr, March 25, 1913, cited in Rosenfeld 1963, p. xiv). This is not the end of the story. I shall, however, pause my account of it here and return to it in closing this chapter.

While Rutherford was primarily an experimental physicist, who also made important theoretical contributions, Bohr was a theoretical physicist, who had, however, done important experimental physics earlier in his career. Bohr's first published paper was on the experiments he had performed himself dealing with the surface tension of liquids, admittedly his only such paper, but a significant experimental contribution, nevertheless. His second published paper dealt with the theoretical part of the same problem and was purely theoretical. These papers stemmed from Bohr's entry into a 1905 prize competition concerning this problem, a competition that Bohr won (Pais 1991, pp. 101–102). Bohr also worked in Rutherford's lab, where he started to develop his ideas concerning the atomic constitution of matter, eventually presented in his 1913 atomic theory. Bohr valued experimental physics greatly and championed its significance throughout his life. According to Heisenberg: "Bohr was primarily a philosopher, not a physicist, but he understood that natural philosophy in our day and age carries weight only if its every detail can be subjected to the inexorable test of experiment" (Heisenberg 1967, p. 95). One might question the view that Bohr was primarily a philosopher, rather than a physicist. In my view, he was both, and I would argue that he was at his best as a philosopher when thinking philosophically about physics, rather than in extending, admittedly in a preliminary and tentative way, his ideas, such as complementarity, beyond physics (the subject that I shall address in Chap. 9). Most crucial at the moment is that Bohr understood that not only theoretical concepts and arguments, but also experimental evidence and arguments have philosophical dimensions to them, dimensions that, moreover, acquire a special significance in quantum theory.

While the full measure of this significance became apparent only after the discovery of quantum mechanics, Bohr's 1913 paper and the earlier work of Planck and Einstein, just mentioned, were harbingers of these complexities, new to physics, although relativity already posed some philosophical questions similar to those posed by quantum theory. I would argue that some of the "German" long-windedness of Bohr's paper, resisted by Rutherford, reflected Bohr's struggle with these complexities and his emerging sense that they might ultimately prove to be unavoidable in quantum theory, and in particular that they could not be handled by means of the kind of thinking that defined classical physics and even relativity.

By contrast, although in turn aware of these complexities, Rutherford was not ready to give up on classical thinking in physics, with which Bohr's most radical moves in the paper were already in conflict, albeit not quite to the degree his later thinking was. Rutherford saw these complexities in classical terms, just as did Einstein, even after quantum mechanics, which, by contrast, brought Bohr to his ultimate understanding of this "entirely new situation" in physics (Bohr 1935, p. 700).

Such complexities can be and often are bypassed in the disciplinary practice of physics, essentially because of its mathematical character, which has defined modern physics since Galileo, who established physics as, in his words, a mathematical science of nature. Quantum theory is no exception, radically different as it may be from classical physics epistemologically; and theoretical physicists can, and most do, productively work on the mathematics of quantum theory and relate this mathematics to experimental data, without engaging with epistemological aspects of the theory. However, these complexities do come into play in deeper foundational questions, such as those that were at stake in Bohr's paper, especially, again, when such questions are precipitated by a crisis. Understanding the quantum aspects of matter could not, in Bohr's view, bypass philosophical issues, in particular those at stake in the long standing and still continuing, and, as I said, perhaps interminable, debate concerning quantum theory. This debate was new then, indeed was it not quite a debate, but rather a problem that was expected by most to be solved on classical-like lines. It took its modern shape with the creation of quantum mechanics in 1925, and Bohr's confrontation with Einstein, which ensued in its wake.

It may be argued, however, that this debate, especially Bohr's confrontation with Einstein, and even Bohr's exchange with Rutherford concerning his article, is a continuation of a much older debate, which extends from the pre-Socratics and, in the modern age, from Descartes on. It concerns the role of philosophical thinking in physics, experimental and theoretical, for example and in particular, as concerns the empiricist and positivist view of physics (especially along Machian lines) versus, first, the realist view, such as that of Einstein, and second, the non realist view, such as that of Bohr, in some respects, already apparent in 1913. Importantly, however, a non realist view need not be positivist or empiricist, and as I explained in the Introduction, that of Bohr was not. The Bohr–Einstein debate is a confrontation between two fundamentally different philosophies of physics and nature, and of our interaction with nature by means of experimental and theoretical physics—a realist and an anti-realist philosophy (keeping the complexities of each, such as those signaled by a possible departure from positivism or empiricism in either case). The still ongoing debate concerning quantum theory continuously replays this confrontation, and brings various versions of these two philosophies and other philosophical positions, such as empiricism and positivism, to the stage of this debate. By referring to "the German method, where it appears to be a virtue to be as long-winded as possible," Rutherford also appears to have referred to a more philosophical method in physics. If so (it is difficult to be certain), Bohr, who had a strong philosophical background and interests, was, even at the time, likely to have had a different assessment of this method and of the pertinence of philosophy in physics in general.[1] This is why I said that Bohr might have taken the title "philosophical magazine" seriously, or at least might have been aware of the ironies involved. The question is to what degree a (more) philosophical way of thinking could or should be brought into physics, or conversely exiled from it, as Rutherford would perhaps have preferred it to be.

It is not that Rutherford did not understand or was hostile to Bohr's argument, at least Bohr's physical argument; quite the contrary, he clearly saw its significance and radical nature, although Bohr's most radical ideas troubled him. Indeed, he remained cautious as to how definite Bohr's argument was for quite some time, expecting a more classical solution of the problem of the atomic constitution, a hope that became even more frustrated with the subsequent developments of quantum theory and made even less likely to be realized by quantum mechanics. It is not clear whether Rutherford ever really reconciled himself to the kind of thinking in physics that Bohr and then Heisenberg adopted. As Rutherford's letter makes clear, he also realized Bohr's desire for and even obsession with clarity and precision, a hallmark of all of Bohr's writings. Rutherford's position appears to have been essentially empiricist or positivist philosophically, and from this position, he may not have perceived the relevance of Bohr's philosophical thinking creeping into his paper (at this stage it was no more than that) to his physical argument. For Rutherford, philosophical thinking was not sufficiently significant or even relevant for physics—it was only part of the long-windedness of the German method. For Bohr philosophical thinking was essential to physics, especially for quantum physics, and his 1913 papers already began to reflect his more philosophical style of thinking in physics. Bohr's most radical, revolutionary epistemological move—the impossibility of offering an ultimate explanation of some physical processes in the atoms—was, I would contend, a product of this fusion of physics and philosophy in his thinking. While, as I said, this move gave Rutherford a pause, later on it inspired Heisenberg and helped his discovery of quantum mechanics, which placed all physical processes inside atoms beyond the reach of explanation.

Bohr's theory of atomic constitution ambitiously aimed to remedy the difficulties of Rutherford's own earlier "planetary model" of the atom, with electrons orbiting atomic nuclei. Although a remarkable conception in turn, this model was inconsistent with classical electrodynamics, which would dictate that the electrons would nearly instantly spiral down into the nucleus, and hence that atoms could not be stable, while they are in fact manifestly stable. Bohr's theory was based on Planck's and Einstein's theories, which, as I explained in Chap. 1, postulated the possibility, in certain circumstances, of the discontinuous emission of light in the form of light quanta (or energy), $h\nu$, eventually understood as photons, particles of light. Making a revolutionary and audacious move, Bohr postulated both the so-called stationary states of electrons in the atom, at which they could remain in orbital motion, and discontinuous "quantum jumps" between stationary states,

---

[1] On Bohr's philosophical background in general, see (Favrhold 1992).

resulting in the emission of Planck's quanta of radiation, without electrons radiating continuously while remaining in orbit, thus, conflicting with classical electrodynamics. In addition, again, in contradiction to the laws of classical electrodynamics, Bohr postulated that there would exist a lowest energy level at which electrons would not radiate, but would only absorb, energy. Bohr also abandoned as hopeless an attempt to offer a mechanical explanation for such transitions, as opposed to the stationary states themselves. The latter, he said, "can be discussed by help of the ordinary mechanics, while the passing of the system between different stationary states cannot be treated on that basis" (Bohr 1913, p. 7). Bohr's postulates were, thus, in manifest conflict with both classical mechanics (because they implied that there is no mechanical explanation for "quantum jumps" between orbits or stationary states) and with classical electrodynamics (because of the way in which, in Bohr's theory, electrons would or, in the case of the lowest energy states, would not radiate energy). Bohr's postulates, however, proved to be correct and have remained part of quantum theory ever since, thus proving that classical theory and laws do not apply to the ultimate quantum constitution of matter. The postulates were given a proper mathematical theory with quantum mechanics.

This summary offers a more or less standard distillation of the essence of Bohr's 1913 atomic theory, which brought him a Nobel Prize in 1922. Thirty years later, Einstein commented on Bohr's "miraculous" achievement as follows: "That this insecure and contradictory foundation [of the old quantum theory] was sufficient to enable a man of Bohr's unique instinct and sensitivity to discover the principle laws of the spectral lines and of the electron shells of the atoms, together with their significance for chemistry, appeared to me as a miracle—and appears to me a miracle even today. This is the highest musicality in the sphere of thought" (Einstein 1949a, pp. 42–43; translation modified). Although beautiful and reflecting the magnitude of Bohr's achievement in a way undoubtedly gratifying to Bohr, the comment, I would argue, still reflects Einstein's unease with the "foundations" upon which Bohr and then Heisenberg built their theories. These foundations never became secure or even acceptable as foundations for Einstein (he may no longer have seen them as contradictory), as his overall reflections on the same occasion make clear.

His earlier view of the situation in 1917 in a related context and Rutherford's immediate response are worth briefly commenting upon here. In the letter to Bohr, cited above, Rutherford said (before proceeding to his remark on Bohr's "arrangement of the paper"): "There appears to me one grave difficulty in your hypothesis, which I have no doubt you fully realize, namely, how does electron decide what frequency it is going to vibrate at when it passes from one stationary state to the other? It seems to me that you would have to assume that the electron knows beforehand where it is going to stop" (Letter to Bohr, March 20, 1913, reproduced in "The Rutherford Memorial Lecture," *PWNB* 3, p. 41). Pais, who cites this passage as part of his account of Bohr's work on his paper and reactions to it in his biography of Bohr, comments on this question and on Einstein's related question a bit later. Pais says: "In typical Rutherford style he had gone right to the heart of the matter by raising the issue of cause and effect, of causality: Bohr's theory leaves unanswered not only the question why there are discrete states but also why an individual electron in a higher [energy] state chooses one particular lower state to jump into. In 1917 Einstein would add a related question: How does an individual light-quantum, emitted in an atomic transition, know in which direction to move?" (Pais 1991, p. 153). Leaving the language of "choice" on the part of electrons and photons aside for the moment (I shall return to this subject later in this study), both Rutherford's and Einstein's statements clearly represent the classical (causal and, in the first place, realist) way of thinking, which neither ever gave up and with which Bohr was willing to part. Contrary to Rutherford's view of Bohr's hypothesis as "a grave difficulty," Bohr saw the situation and, hence, his hypothesis as a solution rather than a problem, thus anticipating and inspiring Heisenberg's attitude in his discovery of quantum mechanics, which answered these questions more fully, albeit not to Rutherford's or, especially, Einstein's satisfaction. Pais concludes: "These questions [of Rutherford and Einstein] were to remain unresolved until … quantum mechanics gave the surprising answer: they are meaningless" (Pais 1991, p. 153). Not to Rutherford and Einstein, or many others following them! Accepting this answer requires a very different epistemological attitude, which is still uncommon and, if accepted as unavoidable, is often seen as an unfortunate and, hopefully, temporary imperative.

Bohr's paper and his "1913 trilogy" of papers devoted to his theory contain further inklings of this epistemological attitude. The limits in this study do not permit me to offer a proper discussion of Bohr's paper. However, Bohr's elaboration reflecting the situation just discussed merits further attention in the context of Heisenberg's discovery of quantum mechanics, to be discussed in the next chapter. Bohr says: "While there obviously can be no question of a mechanical foundation of the calculation given in this paper, it is, however[,] possible to give a very simple interpretation of the result of the calculation on p. 5 [concerning stationary states] by help of *symbols* taken from the mechanics" (Bohr 1913, p. 15; emphasis added). The sentence is best known for its first part, "there obviously can be no question of a mechanical foundation of the calculation given in this paper," which occasioned Rutherford's comment cited above, since it poses, quite dramatically, the question of causality. It is this radical approach that inspired Heisenberg, who echoes this statement when he says: "a geometrical interpretation of such quantum-theoretical phase relations by analogy with those of classical theory seems at present scarcely possible" (Heisenberg 1925, p. 265). Heisenberg's remark is, however, also a reflection on Bohr's sentence as whole, and his approach is a full-scale (rather than limited, as in Bohr's theory) enactment of the program *implicit* in this sentence, even if Bohr himself might not have fully realized these implications or their scale at the time.

We no longer really read Bohr's paper (and few have ever done so) by thinking through each sentence of it, and Bohr, again, always invested a major effort in each of his sentences. I suspect, however, that Heisenberg had read his paper carefully, even though his many discussions with Bohr before and during his work on quantum mechanics would have been sufficient for Heisenberg to know Bohr's thinking concerning the subject and to inspire him. As I shall explain, Heisenberg's own approach in his creation of quantum mechanics in effect amounts to taking "*symbols...* from the ordinary mechanics," where they represent classical physical variables (such as position and momentum) and equations connecting these symbols, and giving both a totally different mathematical form and a new physical meaning. In Heisenberg's theory, these symbols become infinite matrices, infinite square tables of quantities, with a proper rule of algebraically manipulating them, instead of regular functions of coordinates and time, as in classical physics. Physically, these new variables are linked to the probabilities of the occurrences of certain observable phenomena, in this case, spectra, instead of describing the motion of quantum objects on the model of classical mechanics. In this sense, Heisenberg's mechanics was *symbolic* mechanics, as Bohr often referred to it, thus echoing his earlier thinking concerning his 1913 atomic theory and in effect extending it to his interpretation of quantum mechanics, and to his philosophy of physics and nature, a philosophy that quantum theory made imperative for him. Heisenberg's approach reflects the epistemology that made Rutherford's and Einstein's questions, stated above, meaningless, at least to Bohr and Heisenberg, and those (not very many) who were willing to follow them.

I am now ready to finish my Bohr and Rutherford story. Upon receiving Rutherford's letter, cited earlier, which urged Bohr to follow "the custom in England to put things very shortly and tersely, in contrast to the German method, where it appears to be a virtue to be as long-winded as possible," Bohr took a ship from Copenhagen to Manchester. As he recollected in his Rutherford Memorial Lecture after noting the "embarrassing" nature of the situation:

> I therefore felt the only way to strengthen matters was to get at once to Manchester and talk it all over with Rutherford himself. Although Rutherford was as busy as ever, he showed an almost angelic patience with me, and after discussions through several long evenings, during which he declared he had never thought I should prove so obstinate, he consented to leave all the old and new points in the final paper. Surely, both style and language were essentially improved by Rutherford's help and advice, and I have often had occasion to think how right he was in objecting to the rather complicated presentation and especially to the many repetitions caused by references to previous literature. (*PWNB* 3, p. 42)

Well, perhaps! But then something reflecting the character of Bohr's thinking, both physical and philosophical, would be lost as well, and besides we do not know the details of these negotiations and what Rutherford aimed to cut or change. Be that as it may, both Bohr's determination and Rutherford's patience deserve credit for bringing Bohr's paper to publication. Eventually, Bohr received his Nobel Prize for the work presented there. More importantly, it changed the course of the history of physics and the philosophy of physics. So, sometimes physics and philosophy do become reconciled in practice to the benefit of both and their relationships.

They can also be brought together within a given theory, such as quantum mechanics, although it takes the likes of Bohr or Heisenberg to do so. Bohr's thinking concerning quantum phenomena may be said to be both fundamentally physical and fundamentally philosophical. It is fundamentally physical, and not only theoretical but also experimental, *empirical*, one might even say, because, as Heisenberg said, "its every detail can be subjected to the inexorable test of experiment." However, and this is why Bohr's thinking is fundamentally philosophical, no such test, Bohr argues, is possible apart from our conceptual, philosophical determination of both such tests themselves, and what they tell us about how *nature* makes the outcomes of these tests possible. I underline nature here, because while we can always control the setups of such tests, we can never fully control their outcomes, especially in quantum physics, which leads to the irreducibly probabilistic nature of our predictions concerning even primitive individual quantum processes.

In this respect, Bohr follows, *up to a point*, Einstein, his great philosophical enemy, or rather his greatest *philosophical* enemy and his greatest philosophical friend. I stress philosophical, because personally they were always friends. I would argue, however, that they were philosophical friends as well, not the least by being philosophical enemies, because their confrontation helped to advance their conflicting philosophical views. Einstein, as I said, believed, specifically against Mach's views, that we can only develop, as far as it is humanly possible, a true understanding of the nature of physical reality through a *free* conceptual construction, and not merely on the basis of experience. In commenting on Heisenberg's (more Machian) contention in his paper introducing quantum mechanics to the effect that the paper aimed to deal only with "quantities which in principle are observable" (Heisenberg did not quite follow this principle in his argument itself), Einstein told Heisenberg that our concepts and theories decide what could be observed (Heisenberg 1971, p. 63). Einstein's argument impressed Heisenberg and, in part, guided his work on the uncertainty relations. Einstein's insight is crucial because it leads to a questioning of the uncritical use of the idea of observation, an idea that has been a subject of much discussion throughout the history and philosophy of science. He argued against the empiricist or positivist "philosophical prejudice," which "consists in the belief that facts by themselves can and should yield scientific knowledge without free conceptual construction." He added: "such a misconception is possible only because one does not easily become aware of the free choice of such concepts, which, through success and long usage, appear to be immediately connected with the empirical material" (Einstein 1949a, p. 47).

Bohr follows Einstein insofar as free conceptual constructions, mathematical and philosophical, are seen by Bohr as decisive as well: there could be no quantum mechanics or his interpretation of it otherwise. What else is the concept of complementarity, or in his 1913 theory, the concept of a quantum jump? However, he departs from Einstein insofar as this free conceptual construction is, in his interpretation, no longer in the service of describing the ultimate nature of quantum objects and processes, Einstein's key desideratum for a physical theory. In Bohr's scheme quantum objects and processes are placed beyond the reach of our thought altogether. Our conceptual construction is now only in the service of our predictions concerning the outcomes of observable and constructible quantum experiments, or quantum phenomena, made possible by uncircumventably unconstructible quantum objects. Einstein found this way of thinking about physics "logically possible without contradiction," but "very contrary to his scientific instinct," and this was also, and in the first place, his philosophical instinct (Einstein 1936, p. 375). That may be, and yet it was this type of thinking that, as Einstein, as we have seen, in fact acknowledged in invoking its insecure and contradictory foundation, made Bohr's 1913 thinking "the highest musicality in the sphere of thought." This highest musicality had, I would contend, never left Bohr's thinking about quantum physics. If anything its harmonies had become ever more complex, without losing any of their highest musicality.

Chapter 3
1925. "Atomic Theory and Mechanics": From the Old Quantum Theory to Heisenberg's Quantum Mechanics

3.1 From Bohr to Heisenberg, and from Heisenberg to Bohr

This study follows the development of Bohr's thinking concerning quantum phenomena and quantum mechanics, leading him to his interpretation of both, in part based on his concept of complementarity, introduced in the so-called Como lecture of 1927, which was also his first attempt to offer such an interpretation or, at least, to establish a framework for it. As noted at the outset of this study, this development shows refinements and transformations of Bohr's thinking and argumentation—to the point that it is impossible to speak of a single interpretation of quantum phenomena and quantum mechanics in Bohr's case. The situation is different as concerns the concept of complementarity, which, as a general concept, remained more or less the same throughout this development. On the other hand, the key instantiations of this general concept changed in conjunction with changes in Bohr's interpretations, thus also giving rise to new specific physical concepts defined by these new instantiations.[1]

I shall especially distinguish, as reflecting the most significant stages of Bohr's thinking, three versions of his interpretation:

---

[1] These changes are not always sufficiently taken into account by commentators, although the subject has received more attention in recent years, along with the difference between Bohr's interpretation and "the Copenhagen interpretation." On the latter subject, see for example, Howard 2004; Gomatam 2007. These articles, however, offer interpretations of Bohr that are quite different from the one offered here, and, more problematically, they also do not address the evolution of Bohr's views and speak of Bohr's interpretations as a single interpretation. At the same time, some commentators criticize Bohr's argument and "the spirit of Copenhagen" on the grounds of the lack of a single Copenhagen interpretation and of a single (or fully unified) interpretation in Bohr's case, as does, for example, Mara Beller, as part of her advocacy of David Bohm's hidden-variables approach (Beller 1999). As I have argued previously, Beller's argument is unconvincing; it appears to miss some among the most essential aspects of Bohr's argumentation at all stages (Plotnitsky 2002, pp. 254–255, n.33).

A. Plotnitsky, *Niels Bohr and Complementarity*, SpringerBriefs in Physics, DOI: 10.1007/978-1-4614-4517-3_3, © Arkady Plotnitsky 2013

(1). the Como version, based on the complementarity of space–time coordination and the claim of causality;
(2). the pre-EPR version, based on the complementarity of the space–time description and the application of conservation laws, and associated complementarities; and
(3). the post-EPR version, based on the same set of complementarities and Bohr's two new concepts, "phenomenon" and "atomicity."

These three versions of Bohr's interpretation and the corresponding sets of complementary entities will be discussed respectively in Chaps. 4, 5, and 9 (Chap. 7 will discuss quantum field theory and Chap. 8 the Bohr-EPR exchange). The Como version is especially different from the other two, which may be seen in terms of further refinement rather than significant difference. In addition, the Como version contains a significant problem as concerns its treatment of the question of causality in quantum mechanics. Within less than a year, however, Bohr rethought his Como argument, in part under the impact of his exchanges with Einstein, which continued to shape Bohr's thinking from that point on until roughly 1950. The Como argument is unique in that it was not influenced by Einstein's criticism of quantum mechanics. The argument also did not rely on the double-slit experiment, which was first used by Bohr in his initial exchanges with Einstein in 1927 and which played a major role in Bohr's subsequent arguments and debate with Einstein.

The *invention* of (the concept of) complementarity is unimaginable apart from the discovery of quantum mechanics by Heisenberg and Schrödinger, and apart from Heisenberg's discovery of the uncertainty relations, although Bohr began to think about quantum phenomena and quantum mechanics along related lines before the uncertainty relations. The case of the uncertainty relations is especially famous and is often recounted in the literature. Heisenberg was at Bohr's Institute in Copenhagen at the time, and his progress in his work on the physical meaning of quantum mechanics, which led him to the uncertainty relations, appears to have been helped by Bohr's absence from the Institute at the time. Bohr left for a skiing vacation in Norway after weeks of exhausting discussions about the difficulties of quantum mechanics with Heisenberg. (Schrödinger visited earlier, and that, equally famous, encounter is often recounted in the literature as well.) By the time Bohr returned, Heisenberg had finished his paper and submitted it for publication without Bohr having read it, to some chagrin on Bohr's part. Heisenberg famously added a note in the proofs reflecting some of Bohr's criticism of the paper. The paper had a major impact on Bohr's thinking and helped him to develop the full-fledged concept of complementarity. The invention of this concept enabled Bohr to offer an interpretation of the new theory, which ultimately depended on a different instantiation or set of instantiations of complementarity, actually more closely related to the uncertain relations. Bohr's move away from his Como approach, influenced by Schrödinger's wave mechanics, was also accompanied by his return to the key epistemological ideas of Heisenberg's paper introducing his matrix version.

Accordingly, I thought it helpful to devote this chapter to Bohr's initial response to Heisenberg's paper in his 1925 survey "Atomic Theory and Mechanics" (Bohr 1925, *PWNB* 1, pp. 25–51). It was, thus, written not only before Bohr's introduction of complementarity, but also before Schrödinger's wave mechanics and Heisenberg's discovery of the uncertainty relations, both of which shaped the Como argument. It was, however, written in the immediate wake of Heisenberg's paper introducing his, in Bohr's words, "rational quantum mechanics" and the development of Heisenberg's argument into a cohesive version of the theory in (Born and Jordan 1925), and then in the so-called *"Dreimannarbeit,"* "three-man paper" (Bohr, Jordan, and Heisenberg 1926). Bohr's article, originally given as a lecture at the Scandinavian Mathematical Congress in Copenhagen in August 1925, was in preparation as a survey of the state of atomic theory before Heisenberg's discovery of quantum mechanics, but was modified in view of this discovery and Born and Jordan's work on casting Heisenberg's mechanics into its proper matrix form. The survey makes clear that Bohr was among the first to clearly grasp the nature of Heisenberg's ideas and their far-reaching implications. Judging Heisenberg's "step" to be "probably of fundamental importance," Bohr wrote: "In contrast to ordinary mechanics, the new quantum mechanics does not deal with a space–time description of the motion of atomic particles" (Bohr 1925, *PWNB* 1, p. 48).

Bohr was not unprepared for this eventuality, as is clear from his letter to Heisenberg written in the wake of the collapse of the Bohr-Kramers-Slater (BKS) proposal and shortly before Heisenberg's discovery of quantum mechanics (Letter to Heisenberg, 18 April 1925, Bohr 1972–1999, v. 5, pp. 79–80). The epistemology advanced in Heisenberg's 1925 paper could have led Bohr to his ultimate version of complementarity more directly than was his way to it from the argument given in the Como lecture. The possibility of a view that was closer to Heisenberg's paper was intimated in the Como lecture, where Bohr noted that both particles and waves are "abstractions, their properties on the quantum theory being definable and observable only through their interaction with other systems" (Bohr 1927, *PWNB* 1, p. 57). Bohr, however, did not pursue this line of thought in the way he did after his subsequent exchanges with Einstein. This argumentation was not fully settled by Bohr before his introduction, in the wake of EPR's paper (although not in his reply to EPR), of his concept of phenomenon, which only refers to what is classically observed in measuring instruments. By contrast, an assignment of any properties, even single ones and even at the time of measurement, to quantum objects as such is disallowed, which is close to Heisenberg's 1925 approach. Bohr's later view was epistemologically stronger. Heisenberg's theory did not disallow assigning such properties, but was not concerned with them, as in principle *unobservable*, vis-à-vis using the theory for predicting (probabilistically) what could, in principle, be observed in measuring instruments. Bohr's initial commentary, just cited, suggests that Heisenberg's first paper on quantum mechanics contained most of the ingredients that Bohr needed for his ultimate view, even if not all of these ingredients. First of all, the invention of the concept of complementarity was indebted to Schrödinger's wave theory, on the one hand, and the uncertainty relations, on the other. Bohr's thinking leading to complementarity was also influenced by the transformation theory, which connected Heisenberg's and Schrödinger's formalisms and which was developed in 1927 by Dirac (while in Copenhagen) and independently by Jordan.

3.2 Atomic Theory and Mechanics

As Bohr's survey makes clear, quantum mechanics was born, in Heisenberg's work, out of the difficulties and ultimately the impossibility of bringing the mechanical pictures of the old quantum theory, in particular Bohr's own 1913 atomic theory, discussed in Chap. 2, into an adequate correspondence with experiments. Although a major success, Bohr's theory encountered significant difficulties, progressively accumulating during the decade following its introduction. While the theory continued to be developed, again, with some important successes, in order to address these problems, in the work of Bohr himself, Sommerfeld, Heisenberg, and others, it could not cope with these difficulties, which eventually proved to be insurmountable for the theory. In particular, it became no longer possible to represent stationary states by orbits, as opposed to their numerical characteristics ("quantum numbers"). Eventually, a consistent uniform quantum-mathematical treatment of stationary states (no longer "orbits" or conforming to any classical-like mechanical model) and transitions between them was provided by Born and Jordan (Born and Jordan 1925) and by the three-man paper (Born, Heisenberg, and Jordan 1926).

By the early 1920s, it became clear that some drastic adjustments and possibly a new theory, a theory of a *new type*, might be necessary. Although only in his early twenties then, Heisenberg, as just noted, made several significant contributions to the old quantum theory. After completing his doctorate under Sommerfeld in München, Heisenberg was a postdoc of Born at Göttingen, where he first encountered Bohr, who lectured on quantum theory there and who, upon some discussions with Heisenberg, invited Heisenberg to spend some time in Copenhagen. Heisenberg, thus, made Göttingen and Copenhagen the joint birthplace of quantum mechanics. (Schrödinger, who created his wave version slightly later, hailed from Vienna and was in Zurich at the time.) At some point, Heisenberg realized that he did not need a space–time description of quantum objects to be able to predict the outcomes of the experiments he considered, which were concerned with atomic spectra, and he decided to forgo this description altogether. Although in part motivated by the difficulties of observing the actual motions of quantum objects, in this case electrons, this was a daring move, given that throughout the history of mechanics beginning with Galileo or even earlier, such a description, either more direct or implicitly assumed (as in classical statistical physics), appeared necessary for making such predictions. This step was also nontrivial, given that, in yet another unexpected move, Heisenberg retained the classical *equations* of motion. However, the new variables, also introduced by Heisenberg, his great mathematical innovation, no longer related to motion but only to the probabilities of predictions of the outcomes of experiments. Previously, still thinking in the descriptive terms of classical mechanics, the practitioners of the old quantum theory had tried to adjust classical equations while keeping the same physical variables.

Heisenberg's new theory did not deal *only* with "quantities which in principle are observable," as he famously stated in opening his paper, although it was concerned with *predicting only* such quantities. The theory was concerned, even essentially concerned, with (unobservable) quantum objects. As such, it departed from most forms of empiricism and positivism, such as that of Mach, who argued that physics should be essentially concerned only with what can be observed, a view that, as noted earlier, was strongly criticized by Einstein (Einstein 1949b, p. 47). In any event, while—in contrast to classical mechanics and relativity, or even the old quantum theory (which, as just explained, retained a classical-like description of stationary states)—Heisenberg's theory *did not describe* quantum objects and their behavior (even for stationary sates), the *existence* of quantum objects was essential. Heisenberg's argumentation clearly meant that it was this existence that is responsible for the observable quantities in question, even though the way in which quantum objects exist was not considered by the theory and was perhaps unavailable to any theory. The significance of the independent existence of quantum objects was equally crucial for Bohr. As I argue here, ultimately one might need to speak of quantum objects not only as unobservable but also as indescribable or even inconceivable, or one might think of the existence of certain entities in nature that necessitate this concept of quantum objects. The existence of such entities is meaningful even though there is nothing meaningful that we can say about their independent behavior, and any meaningful statement related to them could only concern effects of their interactions with measuring instruments or equivalent media.

While, then, some aspects of classical physics were retained in the old quantum theory (sometimes, accordingly, called semi-classical), Heisenberg's scheme abandoned the application of the concepts of classical physics to quantum objects altogether, in particular the assumption of a classical description of "orbits," even in the case of stationary states. Thus, Heisenberg's new mechanics was a *strictly quantum* theory, the first such theory. It retained the use of classical physics in two capacities. First, classical physics functioned as a certain limit of the new (quantum) mechanics, since the predictions of both quantum and classical mechanics were assumed to coincide, once one reached certain limits within which one could apply both classical and quantum physics to quantum processes. This assumption, which also implies that there are limits beyond which one cannot apply classical physics to quantum processes, constitutes the content of Bohr's famous correspondence principle, although, as will be seen, the principle, crucial for Heisenberg's discovery, was also given a new, more mathematical meaning in Heisenberg's scheme. Second, classical physics could be—and perhaps (as in Bohr's interpretation) would have to be—used in the description of the measuring instruments involved and hence of the actual outcomes of the experiments considered. This "split" into the classical description at the level of measuring instruments and indescribability at the level of quantum objects was essentially different from the "mix" of a classical (e.g., orbits) and quantum (e.g., quantum jumps) description in considering quantum objects and their behavior in the old quantum theory.

The scope of this study does not extend to a discussion of the history of quantum theory from Bohr's 1913 theory of the hydrogen atom to quantum mechanics, a history usefully traced by Bohr himself in "Atomic Theory and Mechanics" (Bohr 1925). In particular, Bohr offers a nice summary of the state of quantum theory as it stood before Heisenberg's discovery, with Kramers and Heisenberg's paper on dispersion theory representing its most significant preceding achievement and a transitional step to Heisenberg's work. Originally, the survey, in the works for quite a while, appears to have been planned to end with this summary, but in the final version this summary ends the penultimate section, "Insufficiency of Mechanical Pictures." The final section, "The Development of a Rational Quantum Mechanics," is devoted to Heisenberg's discovery and to matrix mechanics, developed by Heisenberg, Born, and Jordan. These titles themselves capture the move from the manifest "insufficiency of mechanical pictures" in the old (*semi*-classical) quantum theory to their abandonment in Heisenberg's new "rational quantum mechanics." Bohr's title of his survey, "Atomic Theory and Mechanics" is deliberately precise as well. At stake are the relationships between atomic theory and mechanics, as a theory of individual quantum processes, from the attempts to develop classical-like mechanics to Heisenberg's quantum mechanics. The latter was a mechanics because it predicted, albeit probabilistically, the outcomes of such processes, even though it does not describe them in the way classical physics does for classical objects. That such predictions of individual quantum processes could only be probabilistic even ideally in this case, in contrast to the ideally exact predictions of classical mechanics, was a crucial point in turn.

Heisenberg argued that quantum mechanics required a "new kinematics," which would represent his new variables, and the development of this kinematics proved to be one of his major innovations. Traditionally, as its etymology indicates, "kinematics" refers to a representation, usually by means of continuous (indeed differential) functions, of the attributes of motion, such as spatial (or temporal) coordinates or velocities of a body. The representations of *dynamic* properties, such as momentum and energy, are dependent on and are functions of kinematical properties. By contrast, Heisenberg's new kinematical and dynamical variables related, again, in general probabilistically, to what is observable in measuring instruments under the impact of quantum objects, rather than representing the attributes of the motion of these objects. The kinematical elements of Heisenberg's theory were no longer functions representing properties of quantum objects or their behavior. These elements were conceived as infinite-dimensional square-tables, matrices, of complex variables, eventually rethought in terms of operators in a Hilbert space, or still more abstract mathematical entities. Their main and particular important new feature was that, while one can define the operation of multiplication for them, in general, they do not commute, that is, the order of their multiplication could affect the outcome, $pq \neq qp$. As Born and Jordan discovered in their paper, these matrices had to be infinite to treat these data consistently (Born and Jordan 1925, p. 291). It also became clear shortly thereafter that these matrices needed to be infinite for the matrix scheme to be consistent with the uncertainty relations for continuous variables, such as those of position and momentum.

In his initial commentary on Heisenberg's paper in his 1925 survey, Bohr said that the "fundamental importance" of Heisenberg's step was in "formulating the problems of the quantum theory in a novel way, by which the difficulties attached to the use of mechanical pictures may, it is hoped, be avoided" (Bohr 1925, *PWNB* 1, p. 48). Ultimately, the difficulties were avoided by abandoning such a use altogether, while the formalism, again, enabled excellent probabilistic predictions concerning the outcome of experiments, and the key physical laws, such as momentum and energy conservation laws, still applied (Bohr 1925, *PWNB* 1, pp. 48, 51). The proof of the latter fact was essential for establishing the new mechanics as a proper physical theory, and, in addition, as an irreducibly proba- bilistic theory. It also carried a great significance specifically for Bohr in view of the BKS proposal of 1924, which, as I explained, implies that energy would only conserve statistically and not strictly in quantum interactions. Bohr elaborates as follows:

> In this theory the attempt is made to transcribe every use of mechanical concepts in a way suited to the nature of the quantum theory, and such that in every stage of the computation only directly observable quantities enter. In contrast to ordinary mechanics, the new mechanics does not deal with a space–time description of the motion of atomic particles. It operates with manifolds of quantities which replace the harmonic oscillating components of the motion and symbolize the possibilities of transitions between stationary states in conformity with the correspondence principle. These quantities satisfy certain relations who take the place of the mechanical equations of motion and the quantization rules [of the old quantum theory]. (Bohr 1925, *PWNB* 1, p. 48; also Bohr 1927, *PWNB* 1, pp. 70– 71)

As noted above, the classical equations of motion are formally retained in these relations, but are applied to matrix variables, used only for the purposes of pre- dictions, and no longer to any variables describing the motion of quantum objects. Hence, these equations are no longer equations of motion, although they convert into equations of motion at the classical limit, in accordance with the correspon- dence principle. The variables in question are the so-called "amplitudes" of probabilities of transitions between different states of an atom. One thus retains the term "amplitude" from the classical theory, but reinterprets it so as to deprive it of a reference to the physical (e.g., oscillating) motion of physical objects, which reference defines the use of the term in classical physics, and instead to relate it to the probabilities in question. These "amplitudes" were *de facto* what are now known as "probabilities amplitudes." The rule itself for moving from these "amplitudes" to probabilities and hence *real numbers* (between zero and one), as postulated by Heisenberg, is equivalent to Born's square moduli rule for complex numbers, while amplitudes themselves are treated as *complex vectors* (Heisenberg

1925, p. 263). Born's rule applies more generally to all quantum–mechanical predictions, rather than only in the case of transitions between stationary states.[2] One must keep in mind Bohr's warnings concerning the use of such terms, given that the term "amplitude" cannot be seen here as corresponding to a physical process of the type from which the term originates (Bohr 1929b, *PWNB* 1, p. 17). It is merely another "symbol" borrowed from classical physics but given a quantum–mechanical physical content. The theory becomes only probabilistically predictive rather than descriptive.

Bohr's emphasis on "individual processes" in "Atomic Theory and Mechanics" and then throughout his work on quantum mechanics is crucial. Given the failure of the old quantum theory in this respect, at stake was the first *mechanics* of quantum processes, that is, a theory *relating to individual processes* at the quantum level. The possibility of relating formalism to individual processes, motions, defines classical *mechanics*, as opposed to classical statistical physics, although we do also use the term (classical) "statistical mechanics," which is due to the role of

---

[2] Born's rule works in the following ingenious way, intimated in Heisenberg's first paper on quantum mechanics. Added *ad hoc,* the procedure is not derived from the rest of the quantum–mechanical formalism; it is justified only by the fact that it works—spectacularly well. The procedure is central to quantum mechanics and deserves a special note here. In calculating the probabilities for the outcome of quantum events, such as that of an electron hitting the screen (in a given area), quantum mechanics first assigns to such an event —via the Schrödinger wave function, or the ψ-function—a "probability amplitude." However, this ψ-function is a complex-valued function: the application of such a function generally yields a complex, rather than real, quantity. A complex number is a number of the form $x + yi$, where x and y are real numbers and i is the imaginary unit equal to the square root of -1. Probabilities, on the other hand, are real numbers greater than zero and less than or equal to one. In quantum mechanics, the probability of an event is derived via the absolute square of the amplitude (technically, via the so-called square modulus $x^2 + y^2$ of a complex quantity), written as $|\psi|^2$, which is always a positive real number. Quantum–mechanical formalism allows us to adjust the wave function so as to make the final outcome of the procedure a positive real number that is less than one, just as probabilities are. The procedure involves further technical details glossed over by my summary, such as that the wave function is a multi-dimensional entity and in dealing with such variables as position and momentum, an infinitely dimensional one, or that $|\psi|^2$ is actually what is called the probability density, which means that we need to take an integral of it to get the probability. These complexities, however, do not affect the key points just made. The procedure allows us to properly assess the probabilities involved in quantum experiments, such as those in the interference-pattern setup in the double-slit experiment (to be discussed in detail in Chap. 6) when two paths are open to an electron (again, keeping in mind the qualifications given above). In these cases, we do not add (as we conventionally do in classical physics or elsewhere) the probabilities for the two alternatives—probabilities that can be established on the basis of the alternative setup of the experiment. If we did, our predictions would be incorrect. Instead, we add the corresponding amplitudes, ψ_1 and ψ_2, then derive the probability by squaring the modulus of the sum, $|\psi_1 + \psi_2|^2$, according to the rule just described. The two "states" or "state vectors," usually designated $|\psi_1 >$ and $|\psi_2 >$, are considered to be in "linear superposition." "State vector" is a better term because a given state vector is not seen by Bohr (who actually does not like even this term), except perhaps in the Como lecture, as corresponding to the physical state of a quantum object, although it is seen in this way by some. It is a mathematical entity allowing one to predict a probability of a physical state, and then, in Bohr's ultimate view, only insofar as it is a physical state of the measuring instruments impacted by quantum objects.

underlying individual mechanical processes in the theory. These processes are, however, too numerous to be tracked down exactly, which in fact reduces the theory to statistical estimates concerning these processes. Heisenberg's quantum mechanics was fundamentally different in that, in contrast to classical mechanics, it did not describe individual quantum processes, but only predicted the outcomes of experiments or events involving such processes. By the same token, in contrast to classical statistical physics, it was a *probabilistic* theory of *individual processes and events*, indeed, as Bohr eventually came to realize, especially events. Given these circumstances, one can only speak of these processes as *individual* insofar as, in each case, such a "process" results in an individual event of measurement, observed only in measuring instruments. The actual physical character of such processes is beyond our conception, and hence it may be not be "individual" in any sense we can give to this term, any more than it may be multiple or anything else, a "process," for example.[3]

3.3 Quantum Mechanics and the Correspondence Principle

The correspondence principle was introduced by Bohr in the context of the old quantum theory, and was one of Bohr's major contributions to both old quantum theory and, eventually, to quantum mechanics. Essentially, the principle states that the predictions of quantum and classical mechanics should coincide in the situations where classical mechanics could also be used for predicting the outcome of quantum experiments—for example, for the large quantum numbers for an electron in an atom, when electrons are far from the nucleus. This, as Bohr often stressed, does not mean that the actual behavior of electrons would now change and become classical, but only that it could be sufficiently approximated classically. Heisenberg explains the situation in his uncertainty-relations paper (Heisenberg 1927, pp. 72–76). By this point (in 1927) quantum mechanics was firmly established in both its matrix and wave versions, and the Dirac-Jordan transformation theory, used in Heisenberg's paper, allowed one to use both versions in the same framework. Quantum mechanics, beginning with Heisenberg's first paper on it, also gave a new form to the principle itself, which emerged by virtue of the fact that Heisenberg, in his *new kinematics*, formally adopted the equations of classical physics while changing the variables to which the equations applied. The correspondence principle was now defined by the formal, mathematical *correspondence* between the equations of classical and quantum mechanics. One may speak of this form of the principle as "mathematical," in contradistinction to the previous use of

---

[3] As I shall discuss in Chap. 7, in the case of quantum field theory, it becomes especially difficult to think of such processes as individual, even though an outcome of each such process is always an individual event and is indeed unique each time, just as it is in quantum mechanics.

the principle, which was more physically oriented, insofar as the old quantum theory tried to work with the same type of variables. The physical correspondence in the sense of the identity of both classical and quantum predictions when the classical treatment applies remained in place, but it was established differently, by changing the mathematical nature of variables rather than adjusting equations by using the same type of variables.

This "mathematical" reinterpretation of the correspondence principle was decisive for Heisenberg's work and to the development of quantum mechanics in general. Some, Born, in particular, saw this modification itself as a move beyond the correspondence principle. Born and Jordan, in their first paper on matrix mechanics, claimed that their aim was "to build up the entire [matrix] theory self-independently, without invoking assistance from classical theory on the basis of the principle of correspondence" (Born and Jordan 1925, p. 297; cited in *Mehra and Rechenberg* 2001, v. 3, p. 80). There was an argument concerning the correspondence principle between Born and Heisenberg in writing the three-man paper (Born, Heisenberg, and Jordan 1926). Born contended that, unlike the old quantum theory, the derivation of matrix mechanics was not based on the correspondence principle. Heisenberg argued otherwise, even though the way the principle was now used was very different. According to this paper (in a statement apparently written by Heisenberg), "the new theory ... can itself be regarded as *an exact formulation* of Bohr's correspondence considerations" (Born, Heisenberg, and Jordan 1926, p. 322; emphasis added). In this view, the correspondence principle is seen as "embodying" certain *tendencies* or *principles*, which must be given an "exact," precise formulation by means of a rigorous quantum theory. Heisenberg's new mechanics achieved this formulation, as Bohr says in "Atomic Theory and Mechanics": "The whole apparatus of the quantum mechanics can be regarded as *a precise formulation* of the tendencies embodied in the correspondence principle" (Bohr 1925, *PWNB* 1, p. 49; emphasis added). The parallel between both formulations, highlighted by my emphasis, is striking, although "reformulation" might have been more accurate. Dirac was closer to the point when in his first paper on quantum mechanics he spoke, more cautiously, of the use of (in my terms) mathematical correspondence as "exhausting the content of the correspondence principle" (Dirac 1925, p. 315). He goes so far as to say: *The correspondence between the quantum and classical theories lies not so much in the limiting agreement when $h \Rightarrow 0$ as in the fact that the mathematical operations on the two theories obey in many cases the same [formal] laws* (Dirac 1925, p. 315). Accordingly, one starts by postulating a formal structure, defined by classical (in Dirac's case, Hamiltonian) equations and finds variables that enable correct mathematical predictions in the regions where those of classical theory or even of the old quantum theory fail. Dirac's work was a powerful enactment of Heisenberg's program of using the equations of classical mechanics, while introducing new quantum variables, in his case the so-called q-numbers, which were more general than matrices. As will be seen in Chap. 7, Dirac used the same type of mathematical correspondence in deriving his relativistic equations of the electron, although in this case the correspondence was between relativistic quantum electrodynamics and quantum mechanics, rather than between quantum mechanics and classical mechanics in Heisenberg's case.

From "Atomic Theory and Mechanics" on, Bohr sees the correspondence principle in these terms as well. To be exact, the correspondence principle now functions for Bohr in two ways. The first, which I am primarily addressing here, is mathematical. The second is a related but more physical and subtle use of it, which might also give the principle its most rigorous physical meaning. I shall comment on this second use of the correspondence principle in Chap. 9 and shall merely state here the core point of this use, which is related to the so-called cut [*Schnitt*] between the two domains, where each description, quantum and classical, would respectively apply. This cut is sometimes known as the Heisenberg cut, or the von Neumann-Heisenberg cut in view of its use by von Neumann, which made the idea prominent (e.g., von Neumann 1932, pp. 418–420). The cut is to some degree arbitrary, but only to some degree, since it can only be made, in Bohr's words, "within a region where the quantum–mechanical description of the processes concerned is *effectively* equivalent to the classical description" (Bohr 1935, p. 701; emphasis added). I emphasize "effectively" because, as I said, this equivalence does not mean that these processes themselves are classical. The cut cannot be made above certain limits, in effect defined by Planck's constant h and the uncertainty relations. Below these limits the treatment (technically, it is not a description) of physical processes is strictly quantum.

3.4 Quantum Mechanics and the Relationships Between Mathematics and Physics

Bohr closes his article with a kind of postscript on the role of mathematics in quantum mechanics, a role that he saw as significant as it ever was in *modern physics*, from Galileo on, and on the new type of relationships between mathematics and physics that make quantum mechanics depart from *all physics* hitherto. His comments might be unexpected, given the subsequent trajectory of his thought, especially his insistence on the defining role of measurement rather than on the central significance of mathematics in quantum mechanics. The measuring instruments came to replace "the mathematical instruments," which he invokes here, in his interpretation of quantum phenomena and quantum mechanics from the Como lecture on. In 1925, however, in the immediate wake of Heisenberg's discovery, Bohr writes:

> It will interest mathematical circles that the mathematical instruments created by the higher algebra play an essential part in the rational formulation of the new quantum mechanics. Thus, the general proofs of the conservation theorems in Heisenberg's theory carried out by Born and Jordan are based on the use of the theory of matrices, which go back to Cayley and were developed especially by Hermite. It is to be hoped that a new era of mutual stimulation of mechanics and mathematics commenced. To the physicists it will at first seem deplorable that in atomic problems we have apparently met with such a

limitation of our usual means of visualization. This regret will, however, have to give way to thankfulness that mathematics in this field, too, presents us with the tools to prepare the way for further progress. (Bohr 1925, *PWNB* 1, p. 51)

Bohr was far too optimistic as concerns the physicists' attitudes. While there has been much "thankfulness that mathematics in this field, too, presents us with the tools to prepare the way for further progress," the discontent in question has never subsided and is still with us now. Einstein led the way. He did not find satisfactory or even acceptable either this state of affairs as concerns physics or this type of use of mathematics in physics. Schrödinger was quick to join, with many, even a substantial majority of, physicists and philosophers to follow. However one views the situation on that score, one has to appreciate Bohr's carefulness and precision here, beginning with stressing the *essential* nature of the mathematics in question for quantum mechanics, especially for the rigorous proof of the conservation theorems. Bohr favors the use of the word "essential" in the direct sense of pertaining to the *essence* of a given situation, in other words, as something *fundamental*. It is also significant that Bohr specifically speaks of "a new era of mutual stimulation of *mechanics* and mathematics," rather than physics and mathematics, although Heisenberg's discovery redefines the relationships between them as well. At stake, as I have stressed, are *individual* quantum processes and events. The mathematical science of these processes in classical physics is classical mechanics, which is both descriptive and (actually as a consequence) predictive. It is, correspondingly, classical mechanics that is replaced by quantum mechanics, as a theory that only predicts, in probabilistic terms, such events as effects of certain processes, without describing these processes.

Most crucial are the new *algebraic* relationships in the absence of visualizable and, hence, also geometrical description at the quantum level. It is this situation that defines the "new era" in question. Mathematics becomes even more important for physics. It allows us to predict the outcomes of quantum experiments in the absence of any knowledge or even conception concerning the actual (independent) physical behavior of quantum objects themselves. By the same token, the new relationships between mathematics and mechanics ensure the compatibility between the disciplinary requirements of physics and the new epistemological character of quantum mechanics, from Heisenberg's first paper on it to Bohr's ultimate interpretation of the theory, where this epistemology takes its most radical form. It also follows, not without some irony, that this epistemology opens greater possibilities for the use of mathematical concepts in physics. This is because one's choice of a mathematical scheme under these conditions becomes relatively arbitrary insofar as one need not provide any descriptive physical justification for it, but only needs to justify it by its capacity to make proper predictions. It is true that the actual developments of the mathematical formalism of quantum mechanics emerged (via the correspondence principle) from the descriptively justified formalism of classical mechanics. One can, however, also start directly with this formalism, as Dirac and Born and Jordan did. This is an important part of the new way of "mutual stimulation of mechanics [or physics in general] and mathematics."

As Bohr said later in "Mathematics and Natural Philosophy": "For anyone who through the years has been concerned with the difficulties and paradoxes in quantum physics, it is indeed a deep satisfaction that logical order should be attained to such degree by means of the subtle methods offered by mathematical science" (Bohr 1956, *PWNB* 4, p. 169). Much of this article is devoted to his interpretation of quantum mechanics and the role of measuring instruments there. Bohr's argument is, thus, motivated by the new type of connections between measuring instruments and mathematical instruments in quantum physics, including their jointly resolving the difficulties and paradoxes in question. In his Chicago lectures, Heisenberg argued that "it is not surprising that our language [or concepts] should be incapable of describing the processes occurring within atoms, for … it was invented to describe the experiences of daily life, and these consist only of processes involving exceedingly large numbers of atoms … Fortunately, mathematics is not subject to this limitation, and it has been possible to invent a mathematical scheme—the quantum theory [quantum mechanics]—which seems entirely adequate for the treatment of atomic processes" (Heisenberg 1930, p. 11).

Quantum mechanics, then, appears to be able to relate to that which is beyond language and even thought itself by virtue of its mathematical nature, not subject to the limitations of daily language and concepts, at least those concepts that correspond to the experiences of daily life, from which, as Heisenberg implies here, the concepts of classical physics derive. However, the sense, strictly pre-dictive and not descriptive, in which it does so and in which it is *"entirely ade-quate* for the treatment of atomic processes" also reflects "an entirely new situation as regards the description of physical phenomena," as Bohr later called it (Bohr 1935, p. 700). Not everyone would agree with this claim, beginning, again, with Einstein, who, throughout his life, argued that this scheme is not entirely adequate, an argument that defined his position in his debate with Bohr, which, thus, was also a debate concerning the role of mathematics in physics.

From where Bohr and Heisenberg stand, mathematics now becomes in a certain sense primary, even though quantum mechanics cannot be reduced to mathematics either and, as against classical physics, it contains an irreducible nonmathematical remainder, since no mathematics can apply to quantum objects and processes themselves. But then, nothing else, physics or philosophy, for example, can apply either. The key intuition was that there could be no physical intuition (*Ans-chaulichkeit*), especially in the sense of intuitive (geometrical) visualization of the type found in classical mechanics that could possibly apply to quantum processes, as against using mathematics, essentially algebra, to predict the outcome of experiments. German *"Anschaulichkeit"* conveys this sense better that English "intuition," compelling Bohr to primarily use "visualization" in English in this context. (Terms such as pictorial visualization, pictorial representation, and mental pictures were used by Bohr as well.) The impossibility of such visualization becomes one of the central themes of Bohr's work, beginning with the essays collected in *PWNB* 1 (e.g., Bohr 1929, *PWNB* 1, pp. 98–100, 108). Ultimately, quantum objects and processes are placed beyond the reach of any representation, intuition, or conception. Instead of the geometrical intuition of visualization of classical physics, the quantum–mechanical situation required the kind of algebraic physical–mathematical intuition displayed by Heisenberg. This intuition, however, while not strictly mathematical, depends fundamentally on the role of mathematics, even as it redefines the relationships between mathematics and physics, and transforms the nature of theoretical physics, a transformation that I shall further consider in Chap. 7.

Bohr's elaboration under discussion reflects his profound understanding of this situation. It may appear to announce a program that is more Heisenbergian than Bohrian and that is different from the one Bohr came to follow later. However, by taking this view one would underestimate the subtler complexities of Bohr's thinking concerning the significance of mathematics in quantum mechanics. Instead, from "Atomic Theory and Mechanics" on, Bohr's views are defined by the essential roles of both measuring and mathematical *instruments* in their reciprocal relationships in quantum physics. The very appeal to "instruments" is hardly casual. Apart from the fact that such choices of expression are rarely casual in Bohr, the point is consistent with Bohr's general view of mathematics and its role in physics, as expressed, in particular, in "Mathematics and Natural Philosophy," which makes this reciprocity of both kinds of instruments especially apparent (Bohr 1956, *PWNB* 4). It is, however, manifest throughout Bohr's writings, thus confirming Heidegger's insight that, from Galileo on, all modern physics (classical, relativistic, and quantum) is mathematical because it is experimental and experimental because it is mathematical (Heidegger 1967, p. 93). With quantum mechanics, however, this mutually reciprocal determination of mathematics and experiment in physics takes an entirely new form, with which, in this respect Bohr's hope proved to be justified, "a new era of mutual stimulation of mechanics and mathematics commenced." This era has by now extended for nearly a century and may continue for a long time to come, eager as many, beginning again, with Einstein, have been for this era to come to an end. We have of course also been accustomed to be surprised by nature, and it may surprise us again, perhaps by bringing us closer to fulfilling Einstein's hope. But that it will take us even further away from this hope than quantum mechanics has is not inconceivable either.

Chapter 4
1927. "The Quantum Postulate and the Recent Development of Atomic Theory" (The Como Lecture): Complementarity Versus Causality

4.1 From Quantum Mechanics to Complementarity

This and the following chapter consider Bohr's new way of thinking about quantum phenomena and quantum theory, in terms of his concept of complementarity, a way of thinking that emerged following and in response to the discovery of quantum mechanics and then the uncertainty relations. Bohr's thinking, I argue here, underwent several changes even at the earlier stages to be discussed in these two chapters, and it was further refined in the wake of EPR's argument. The first change is the shift from Bohr's pre-complementarity view of quantum theory in the wake of Heisenberg's 1925 discovery of quantum mechanics, as discussed in the preceding chapter, to his view, via complementarity, following Schrödinger's wave mechanics, Dirac's and Jordan's transformations theory, and Heisenberg's uncertainty relations. These developments of quantum mechanics followed in quick succession and were instrumental to the invention of the concept of complementarity, introduced in the Como lecture in 1927. This change will be discussed in this chapter. The next change, discussed in the next chapter, was marked by Bohr's rethinking of the quantum situation and occurred under the impact of his exchanges with Einstein in 1927, specifically at the Fifth Physical Conference of the Solvay Institute in October 1927 in Brussels. Yet one more major change in Bohr's thinking, leading him to his more or less final version of his interpretation of quantum phenomena and quantum mechanics occurs under the impact of EPR's paper and related arguments by Einstein, to be discussed in Chaps. 8 and 9.

In this chapter, I proceed as follows. This section gives an introductory discussion of the developments of Bohr's concept of complementarity and his initial interpretation of quantum mechanics. Sections 4.2 and 4.3 offer an analysis of the Como lecture. Section 4.3 also examines the problematic aspects of the Como argument, which are linked to the question of causality. As the persistent use of the term "causality" in his subsequent essays indicates, this question was to preoccupy Bohr for the remainder of his life, but it is given a very different treatment in his later works than in the Como lecture.

A. Plotnitsky, *Niels Bohr and Complementarity*, SpringerBriefs in Physics,
DOI: 10.1007/978-1-4614-4517-3_4, © Arkady Plotnitsky 2013

By the "Como argument," the "Como version" of complementarity, the "Como interpretation" of quantum mechanics, and so forth, I refer to the corresponding aspects of Bohr's article, "The Quantum Postulate and the Recent Development of Atomic Theory," based on his lecture given in Como, but published in this version in 1928 (Bohr 1927, PWNB 1, pp. 52–91). As is well known, the article deviates from the lecture originally given in Como, at the conference occasioned by the centenary of Alessandro Volta's death. It has a complicated history of writing and publication, and the text appears to have never been quite finalized by Bohr to his satisfaction. Almost nothing ever appears to have been, at least prior to his late articles, written in the 1950s (which, however, more or less present previously developed arguments). In part, this incompleteness reflects the general character of Bohr's writings, which require one to consult Bohr's other articles to "complete" Bohr's argument, and it is symptomatic that Bohr never wrote a book offering a sustained presentation of his interpretation, in any of its versions. In the case of the Como lecture, however, the situation is especially complicated, which has led to a considerable debate concerning the published version, as regards its correspondence to the lecture itself, possibly better reflected in the preserved earlier drafts, the relative merits and the relationships between these drafts themselves, and so forth. It would, accordingly, be difficult to argue that the published version is the definitive statement of Bohr's views at the time. However, it appears to be at least as definitive as any available versions of the Como lectures to be treated as representing at least a certain stage of his views. It was so treated by Bohr himself, for example, in his "Introductory Survey" to *"Atomic Theory and the Description of Nature"* (PWNB 1, pp. 9–15), originally published in 1931, although, as will be seen in Chap. 5, the survey presents the Como argument with some inflections corresponding to his later views. This collection gives 1927, the date of the lecture itself, as the article's date, although this version was not published until 1928. In any event, the fact that Bohr reprinted the article in this form in this collection suggests that he saw it as a stage of his thinking concerning quantum phenomena and quantum mechanics, a stage especially important by virtue of being defined by the invention of complementarity.[1]

Bohr's debate with Einstein, which had decisively shaped, and vis-à-vis the Como argument *reshaped*, Bohr's thinking, began in the immediate wake of the Como lecture and had continued into 1940s, although Bohr's final version of complementarity and interpretation of quantum mechanics were more or less established by the late 1930s, with only minor nuances added later. It is intriguing, although difficult, to speculate how Bohr's thinking would have developed apart

---

[1] For the history and various versions and drafts of the lecture and the article and commentary on them by J. Kalckar, see volume 6 of Bohr's collected works (Bohr 1972–1999, v. 6). For a recent challenge to Kalckar's account, see (Petruccioli 2011). While the examination of Bohr's various drafts sheds additional light on the genesis of complementarity and helps to ascertain with greater accuracy when certain key ideas actually emerged, it would not, for the reasons just stated, significantly affect the present argument. Most of these drafts contain the key aspects of Bohr's argument found in the published version and reflect the same problems.

from Einstein's arguments. The Como argument was the only one not subject to this pressure. The transition from the Como version to the later versions of Bohr's interpretation is, first of all, marked by a shift from a more philosophical to a more empirical approach to the concept of complementarity, or more accurately, a shift in balance between philosophical and empirical aspects in the corresponding instantiations of the concept. The architecture of the concept was not quite established in the Como lecture either and was only rigorously worked out subsequently. The complementary features or phenomena considered by Bohr change in accordance with this shift as well. The defining complementarity of the Como lecture is that between space–time coordination and the claim of causality, accompanied by some additional features, discussed below, and this complementarity was abandoned by Bohr shortly thereafter. Bohr shifts to the complementarity of certain observable quantum phenomena, or correlatively, the complementarities between the applications of the space–time description and conservation laws (Bohr actually introduced the second type of complementarity first).

Eventually, such complementary phenomena, and all quantum phenomena, were rethought in terms of Bohr's concepts of phenomenon and atomicity, which became nearly as significant for him as the concept of complementarity and in particular made his epistemology even more radical. Bohr's epistemology prior, roughly, to 1937 allows one to associate certain physical properties with quantum objects, if only at the time of measurement. By virtue of the uncertainty relations, however, at most only a half of the properties that define the behavior of classical physical systems could be available for this association at any given point—for example, either a position or a momentum, but never both simultaneously. Bohr appears to have always been ambivalent concerning even this type of association, given that any actual physical quantities observed are only observed in measuring instruments. But, at least after the Como lecture, he also viewed the idea and language of the "creation" of the physical properties of quantum objects by measurements with as much suspicion as those of "disturbance" of (preexisting) properties by measurements (Bohr 1938, PWNB 4, p. 104; Bohr 1949, PWNB 2, p. 64; Bohr 1954, PWNB 2, p. 73). Bohr's concept of phenomenon dispenses with both notions. In Bohr's ultimate view, based on this concept, an association of physical properties with quantum objects is never possible even as concerns single such properties (rather than only of simultaneous conjugate properties, in accordance with the uncertainty relations) at any point—before, during, or after a measurement. All physical properties considered are now only those of certain parts of the measuring instruments involved and are, as physical properties, described by classical physics. The attribution of such properties is still constrained by the uncertainty relations, but now applied to these classical properties. One can never properly define both conjugate classical properties and the corresponding measurable quantities, such as those of position and momentum, for the relevant part of the measuring instrument in the same experimental arrangement. In classical mechanics, one can always do so because one can do so for the corresponding objects themselves, in the first place.

These changes were crucial to the development of Bohr's thinking. It would, however, be equally misleading to disregard the continuities in this development and the significance of the Como argument for it, beginning with the introduction of the concept of complementarity and a conceptual *framework*, the first of its kind, for understanding and interpreting quantum mechanics. Several other contributions in the Como lecture are important as well. I shall comment on them in more detail in closing this chapter, but it is worth briefly summarizing them here. Bohr's title concept of the quantum postulate is crucial to all of his thinking. So is his argument for the irreducible role of measuring instruments in the constitution of quantum phenomena, which intimates Bohr's later concept of phenomenon. This view of measuring instruments leads Bohr to yet another important argument, to the effect that both concepts, particle and wave, are abstractions, derived from the effects of the interactions between quantum entities and measuring instruments upon the latter. As a result, a more rigorous and effective understanding of the wave and the particle aspects of quantum phenomena becomes possible. Bohr's elementary derivation of the uncertainty relations directly from de Broglie's formulas is noteworthy as well. In sum, its contributions make the Como lecture a major achievement in spite of problematic aspects of Bohr's argumentation there.

Arguably, the most crucial of these problems concerns the question of causality. The Como argument attributes causality to independent ("undisturbed") quantum processes, viewed as described by Schrödinger's equation, while the impossibility of deterministic predictions is due to the disturbance introduced by measurements. This concept is abandoned by Bohr, following his exchanges with Einstein shortly after the Como lecture. By the same token, wave mechanics loses its significance in Bohr's post-Como thinking, although the reinterpretation of quantum waves in terms of probabilities and Schrödinger's equation itself retain their fundamental role.

As indicated earlier, however, the significance of Schrödinger's mechanics for the Como lecture does not amount to the significance of wave–particle complementarity. The status of this complementarity is complex even in the Como version, if one can speak of this complementarity in Bohr at all. Bohr not only does not invoke but also does not appear to think in terms of wave–particle complementarity even in the Como lecture, let alone in his subsequent work, although the Como lecture may suggest the idea and appears to have been responsible for its emergence and proliferation. Bohr does relate the wave and the particle features— or, again, more accurately, the wave-*like* and the particle-*like* features—of quantum phenomena to the concept of complementarity in the Como lecture (e.g., Bohr 1927, PWNB1, pp. 55–57) and then elsewhere. These relations, however, do not amount to wave–particle complementarity. There are several reasons why Bohr did not, and why one might not want to, speak in terms of wave–particle complementarity, and I shall consider these reasons below.

As I said, the Como version or instantiation of the concept of complementarity may be seen as more philosophical, while the subsequent versions may be seen as more empirical. In particular, the Como version is defined by the complementarity of space–time coordination and the claim of causality, which, Bohr adds, "[symbolize] the idealization of observation and definition respectively" (Bohr

1927, PWNB 1, p. 55). This qualification gives this complementarity a more philosophical conceptual content by coupling it to the idealizations in question, that of observation and that of definition, although, as I shall explain, the definition in question is primarily mathematical. This complementarity does have an important physical dimension as well by virtue of the fact that, in contrast to classical physics, it splits causality and space–time coordination and, thus, redefines the relationships between experiment (observation) and theory (definition). Once again, there is no invocation of wave–particle complementarity in the lecture. Instead, Bohr relates the wave-like and the particle-like features of quantum phenomena to the complementary of the space–time coordination and the claim of causality (e.g., Bohr 1927, PWNB 1, pp. 55–57).

In any event, the complementarity of space–time coordination and the claim of causality disappears from Bohr's subsequent arguments. Bohr's approach becomes centered, first, on the complementarity of the application of spatial–temporal description and conservation laws, and, correlatively, on the mutual exclusivity of the corresponding experimental arrangements (e.g., those suited for position or, conversely, momentum measurements), linked to Heisenberg's uncertainty relations. The experimental considerations involved in this type of thinking are significant in the Como version as well, but are subordinated there to the complementarity of space–time coordination and the claim of causality. This reorientation defines the key works following the Como lecture, beginning with Bohr's 1929 article "The Quantum of Action and the Description of Nature" (Bohr 1929a), to be discussed in the next chapter. The article, written almost immediately in the wake of Bohr's first exchanges with Einstein on quantum mechanics, represents the transition in question and offers the version of Bohr's interpretation that was in place even in Bohr's reply to EPR. The ultimate version of Bohr's interpretation, grounded in his concepts of phenomenon and atomicity, is developed later on and is first found in "Complementarity and Causality" (Bohr 1937) or in the Warsaw lecture, "The Causality Problem in Atomic Physics" (Bohr 1938).

4.2   The Quantum Postulate: Discontinuity and Irrationality

Bohr appears to have initially formulated the concept of the quantum postulate in the Como lecture, where this concept—and it is, importantly, a *concept*, as well as a postulate—plays a central role, underscored by its appearance in Bohr's title, as the only physical concept there, "The Quantum Postulate and the Recent Development of Atomic Theory."[2] The significance of this concept had never

---

[2] Bohr uses the phrase "the quantum postulates" (in plural) on earlier occasions, including in "Atomic Theory and Mechanics" (Bohr 1925), in referring to basic postulates of his atomic theory of 1913. In the Como lecture, however, the phrase designates a new concept.

diminished in Bohr's work, even as other concepts, beginning with complementarity, moved to center stage. Indeed, Bohr's work on complementarity may be seen as an effort to give, by means of this concept, the proper meaning to "Planck's discovery of the *elementary quantum of action*" (Bohr 1929, PWNB 1, p. 92; emphasis added). Although Bohr often speaks of the postulate as *Planck's* quantum postulate, which is an unavoidable association, his concept of it is his own. "Atomic Theory and Mechanics" (Bohr 1925, PWNB 1, pp. 25–51), discussed in the preceding chapter, does not yet contain this concept. Bohr merely speaks of Planck's discovery as "demand [ing] an element of discontinuity of atomic processes quite foreign to the classical theory" (Bohr 1925, PWNB 1, p. 28). By the time of the Como lecture, however, the quantum postulate is seen by Bohr as expressing the very essence of quantum theory, "characterized by the acknowledgment of a fundamental limitation in the classical physical ideas when applied to atomic phenomena." "The situation thus created," he adds, "is of a peculiar nature, since our interpretation of the experimental material rests essentially upon the classical concepts" (Bohr 1927, PWNB 1, p. 53). These statements reveal the complexity of Bohr's view of the role of the classical concepts as essential and yet limited, and as concerns quantum objects and their behavior, inapplicable, a subject that I shall discuss later in this study.

The quantum postulate itself reflects and, according to Bohr, symbolizes this situation. According to Bohr: "Notwithstanding the difficulties which, hence, are involved in the formulation of the quantum theory, it seems, as we shall see, that its essence may be expressed in the so-called quantum postulate, which attributes to any atomic processes an essential discontinuity, or rather individuality, completely foreign to the classical theories and symbolized by Planck's quantum of action" (Bohr 1927, PWNB 1, p. 53). Bohr's hesitation ("or rather") is worth noting. The essential discontinuity and the essential individuality will eventually be reinterpreted in terms of Bohr's concepts of phenomenon and atomicity, and connected by making this discontinuity or discreteness refer to that between individual phenomena in Bohr's sense. The kernel of the idea is already found here, however. Bohr's appeal to the symbolic nature of Planck's quantum of action, h, is in part due to his view of all quantum-mechanical formalism as symbolic—a view that, as discussed in Chap 3, he developed following Heisenberg's introduction of matrix mechanics and applied to Schrödinger's wave theory as well. This formalism is symbolic by virtue of the fact that, while formally analogous to classical mechanics, it does not describe the physical behavior of quantum objects or any physical processes in space and time, but only serves to predict the outcomes of experiments. Correlatively, Planck's quantum of action is symbolic insofar as it should not be seen as merely implying the discrete, particle-like nature of quantum objects, as opposed to their wave-like character, since both of these aspects are seen as pertaining only to certain effects registered in measuring instruments. Bohr elaborates as follows:

> This postulate implies a renunciation as regards the causal space–time coordination of atomic processes. Indeed our usual description of physical phenomena is based entirely on

the idea that the phenomena concerned may be observed without disturbing them appreciably. This appears, for example, clearly in the theory of relativity, which has been so fruitful for the elucidation of the classical theories. As emphasized by Einstein, every observation or measurement ultimately rests on the coincidence of two independent events at the same space–time point. Just these coincidences will not be affected by any differences which the space–time coordination of different observers otherwise may exhibit. Now, the quantum postulate implies that any observation of atomic phenomena will involve an interaction with [between a quantum object and?] the agency of observation not to be neglected. Accordingly, an independent reality in the ordinary physical sense can neither be ascribed to the phenomenon [object?] nor to the agencies of observation. After all, the concept of observation is in so far arbitrary as it depends upon which objects are included in the system to be observed. Ultimately, every observation can, of course, be reduced to our sense perceptions. The circumstance, however, that in interpreting observations use has always to be made of theoretical notions entails that for every particular case it is a question of convenience at which point the concept of observation involving the quantum postulate with its inherent "irrationality" is brought in. (Bohr 1927, PWNB 1, pp. 53–54)

The last point reflects the essential role of theoretical, including mathematical, concepts in shaping our observation when they enter a physical theory; and, unlike the role of measuring instruments, which is irreducible only in quantum physics, our concepts are irreducible in both quantum and classical physics, or in relativity. The passage contains several ingredients of Bohr's subsequent conceptions and epistemology, although the key distinction between *unobservable* and, ultimately, *inconceivable* quantum objects and *observable* phenomena is not found here. Indeed, as indicated by my questions interpolated into the quotation, there appears to be some confusion in this respect, or at least imprecision in expression on Bohr's part.

An emphasis on—and the very notion of—"disturbance" contributes to the problems of Bohr's argument here. At one level this emphasis is understandable insofar as in the case of quantum phenomena the interference of measurement cannot be avoided and, hence, no independent behavior of quantum objects can be observed. However, the concept of disturbance is problematic because it may imply the possibility of a (specific) conception of the undisturbed behavior of quantum objects. As I said, this concept, too, is abandoned by Bohr in his later work, along with that of "creation" of the attributes of atomic objects by measurements (Bohr 1938, PWNB 4, p. 104; Bohr 1949, PWNB 2, p. 64; Bohr 1954, PWNB 2, p. 73).

On the other hand, I do not find troubling Bohr's appeals to the "inherent 'irrationality'" of the quantum postulate or to the "choice" on the part of atoms or nature later in the article. The notion and the very language of irrationality have often been used against Bohr by his critics and have troubled even his advocates, in my view, as a result of misunderstanding Bohr's thinking.[3] Apart from dissociating quantum mechanics and his own views from any "mysticism foreign to the

---

[3] These concerns and some of this misunderstanding are exhibited, for example, by Kalckar in his introductory commentaries on Bohr's work on complementarity in volumes 6 and 7 of Bohr's collected works, devoted to complementarity (Bohr 1972–1999, v. 6 and v. 7).

spirit of science," Bohr stressed the *rational* character of quantum mechanics throughout his work, beginning with his initial response to Heisenberg's introduction of quantum mechanics (Bohr 1925b, PWNB 1, p. 48; Bohr 1949, PWNB 2, p. 62). The "irrationality" invoked here and elsewhere in his earlier writings is clearly *not* any "irrationality" of *quantum mechanics* itself, which Bohr, again, sees as a *rational* theory, a rational theory of something that may, in a certain sense, be *irrational*—that is, inaccessible to rational thinking or thinking in general. Bohr's point is misunderstood in part by virtue of overlooking (beginning with Bohr's quotation marks) the difference between the rationality of a theory and the irrationality of what this theory rationally deals with. Bohr's invocation of "irrationality" appears to be based on an analogy with irrational numbers, reinforced by the role of complex numbers and specifically the square root of -1, *i* in quantum mechanics. Complex numbers are analogous to irrational numbers, because both are solutions of equations that do not belong to the domain where these equations are defined, the integers in the case of the irrationals, and the real numbers in the case of complex numbers. The ancient Greeks, who discovered the (real) irrationals, could not find an arithmetical, as opposed to geometrical, form of representing them. The Greek terms "alogon" and "areton," which may be rendered as "incommensurable" and "incomprehensible," are equally fitting in referring to quantum objects and processes. The problem of properly defining irrational numbers was only resolved in the nineteenth century, after more than two thousand years of effort. It remains to be seen whether quantum-mechanical "irrationality" will ever be resolved in this way, that is, by discovering a way to mathematically or otherwise represent quantum objects and processes, assuming that one sees it as a problem.

In Bohr's view, the quantum postulate implies the existence of a certain boundary ("cut"), up to a point (but only up to a point) arbitrarily placed, between the "rational"—comprehensible and measurable, or observable and conceivable— classical world and the "irrationality" of the quantum postulate. Quantum mechanics and Bohr's interpretation of it, via complementarity, are rational forms of understanding this situation. The point is brought home in Bohr's 1929 "Introductory Survey" to his *Atomic Theory and the Description of Nature* (PWNB 1): "A conscious resignation in this respect [the impossibility of carrying forward a coherent causal description of atomic phenomena] is already implied in the form, *irrational from the point of view of the classical theories*, of those postulates ... upon which the author based his application of the quantum theory to the problem of atomic structure" (Bohr 1929b, PWNB 1, p. 7; emphasis added). Bohr offers another helpful elaboration there. It follows his comments on complementarity as separating certain features that "are united in the classical mode of description." In so doing, complementarity helps us to avoid the contradiction and "irrationality" (apparently unavoidable if one retains the classical viewpoint). Then he says:

Moreover, the purpose of such a technical term is to avoid, so far as possible, a repetition of the general argument as well as constantly to remind us of the difficulties which, as already mentioned, arise from the fact that all our ordinary verbal expressions bear the stamp of our customary forms of perception, from the point of view of which the existence of the quantum of action is an irrationality. Indeed, in consequence of this state of affairs even words like "to be" and "to know" lose their unambiguous meaning. In this connection, an interesting example of ambiguity in our language is provided by the phrase used to express the failure of the causal mode of description, namely, that one speaks of a free choice on the part of nature. Indeed, properly speaking, such a phrase requires the idea of an external chooser, the existence of which, however, is denied already by the use of the word nature. We here come upon a fundamental feature in the general problem of knowledge, and we must realize that, by the very nature of the matter, we shall always have last recourse to a word picture [picture in words], in which the words themselves are not further analyzed. (Bohr 1929b, PWNB 1, pp. 19–20)

Bohr's "example" is not accidental and has its history beginning, as we have seen in Chap. 2, with the question "How does an electron decide what frequency it is going to vibrate at when it passes from one stationary state to the other?" asked by Rutherford upon reading Bohr's 1913 paper on atomic constitution. While he and others, in particular Dirac (who even spoke of an electron as having a "free will"), used expressions like "a free choice on the part of nature," Bohr's use of it must be considered with the above comment in mind, at least after his 1927 discussions with Einstein at the Solvay conference in Brussels.

Heisenberg offers a penetrating comment in response to Dirac's appeal to "the choice on the part of nature," which Heisenberg questioned on experimental grounds, in the course of the general discussion that took place at the same conference. Dirac's overall comment in effect implied the causal nature of the independent quantum behavior undisturbed by observation in accord with his transformation theory paper (Dirac 1927) and Bohr's Como argument, in this respect influenced by that paper of Dirac. By contrast, while referring to his uncertainty relations paper (Heisenberg 1927), which used the *mathematics* of Dirac's transformation theory, Heisenberg expressed a view that was close to Bohr's post-Como thinking, which no longer assumes that independent, undisturbed quantum processes are causal. This type of view guides Bohr's subsequent analyses of the double-slit experiment, which significantly figured in the Bohr–Einstein exchange and in the general discussion at the Solvay conference. Heisenberg says (according to the available transcript):

I do not agree with Dirac when says that in the [scattering] experiment described nature makes a choice. Even if you place yourself very far from your scattering material and if you measure after a very long time, you can obtain interference by taking two mirrors. If nature had made a choice, it would be difficult to imagine how the interferences are produced. Obviously we say that nature's choice can never be known until the decisive experiment has been done; for this reason we cannot make any real objection to this choice because the expression "nature makes a choice" does not have any physical consequence. I would rather say, as I have done in my latest paper [on the uncertainty relations], that the *observer himself* makes the choice because it is not until the moment when the observation is made that the "choice" becomes a physical reality. (Bohr 1972–1999, v. 6, pp. 105–106)

The technical details of the experiment are not important at the moment; the experiment itself is essentially equivalent to the double-slit experiment, to be discussed in Chap. 6. The crucial point is that we, in effect, deal with the complementary character of certain quantum experiments and our choice of which of the two mutually exclusive or complementary experiments we want to perform, rather than with a choice of nature. Without quite realizing it, Heisenberg describes the so-called delayed choice experiment of John A. Wheeler, to which I shall return in Chap. 6 (Wheeler 1983, pp. 190–192).

Bohr's comment cited above was written in 1929, following the Solvay conference and further exchanges, and it refers specifically to his article, "The Quantum of Action and the Description of Nature" (Bohr 1929), to be discussed in the next chapter. At stake here is not merely a matter of the metaphorical use of such an expression, but, as Heisenberg's remarks make clear as well, a deep epistemological point, which eventually made Bohr avoid speaking in terms of "choice on the part of nature." The situation implies, as Bohr will say in his reply to EPR, "the necessity of a final renunciation of the classical ideal of causality and a radical revision of our attitude towards the problem of physical reality," and ultimately a radical renunciation of the classical ideal of reality as well (Bohr 1935, p. 697). The difference, reflected in Bohr's different phrasings, is that, while causality is renounced at the quantum level altogether, the existence of quantum objects is assumed, without, however, at least in Bohr's ultimate view, our being able to conceive of this existence, even by use of such words as to "be" or to "know." These words, invoked by Bohr in his 1929 elaboration cited above, are not accidental either. They are common in our everyday language but, as Bohr clearly implies, they are complex philosophically, referring to the philosophical problems of ontology and epistemology, which have been the subject to profound philosophical discussions from the pre-Socratics to Heidegger and beyond, and they acquire radically new dimensions with quantum theory. Bohr is known to have replied, after the rise of quantum physics but before quantum mechanics, to Harald Høffding's question "Where can the photon be said to be?" with "To be, to be, what does it mean to be?" (cited in Wheeler 1998, p. 131). Both of these questions are still unanswered and, in Bohr's ultimate view, are unanswerable. Bohr's ultimate ontology allows quantum objects to exist, but it does not and cannot say anything about them apart from the manifest effects of their interactions with the classical world, where such questions as "Where can something be said to be?" can be answered or meaningfully asked, to begin with.

Under this assumption, what is "sometimes picturesquely described as a 'choice of nature' [between different possible outcomes of a quantum experiment]" will be given by Bohr a different meaning as well. As he says, "needless to say, such a phrase implies no allusion to a personification of nature, but simply points to the impossibility of ascertaining on accustomed lines directives for the course of a closed indivisible phenomenon" (Bohr 1954, PWNB 2, p. 73). It points to the impossibility of a causal or any description of quantum processes. This understanding, discussed in Chap. 9, was yet a few years away even in 1929, by which time, however, Bohr managed to overcome the problem of causality that plagued the Como argument.

4.3 Complementarity and the Problem of Causality

The concept of complementarity is introduced in the Como lecture as a consequence of the quantum postulate and the epistemology that it implies, especially the fact of the irreducible "disturbance" of quantum objects by observation. The concept of "disturbance," again, eventually evolved into that of the irreducible role of measuring instruments in the constitution of quantum phenomena. Bohr's preferred term became "interference," which is more precise, because it no longer suggests an assumption of independent quantum processes that are causally describable by quantum–mechanical formalism and that would be "disturbed" by measurement, a defining and problematic assumption of the Como lecture. There, Bohr argues as follows:

> This situation has far-reaching consequences. On one hand, the definition of the state of a physical system, as ordinarily understood (i.e., in classical physics), claims the elimination of all external disturbances. But in that case, according to the quantum postulate, any observation will be impossible, and, above all, the concepts of space and time lose their immediate sense. On the other hand, if in order to make observation possible we permit certain interactions with suitable agencies of measurement, not belonging to the system, an unambiguous definition of the state of the system is naturally no longer possible, and there could be no question of causality in the ordinary sense of the word. The very nature of the quantum theory thus forces us to regard the space–time co-ordination and the claim of causality, the union of which characterizes the classical theories, as complementary but exclusive features of the description, symbolizing the idealization of observation and definition respectively. Just as the relativity theory has taught us that the convenience of distinguishing sharply between space and time rests solely on the smallness of the velocities ordinarily met with compared to the velocity of light, we learn from the quantum theory that the appropriateness of our usual causal space–time description depends entirely on the small value of the quantum of action as compared to the actions involved in ordinary sense perceptions. Indeed, in the description of atomic phenomena, the quantum postulate presented us with the task of developing a "complementarity" theory the consistency of which can be judged only by weighing the possibilities of definition and observation. (Bohr 1927, PWNB 1, pp. 54–55)

The task of developing an effective complementarity theory will take Bohr a while to complete, and it will, I argue, require a way of thinking different from the one that is transpiring here and is developed in the Como lecture, as Bohr realized with Einstein's help shortly thereafter. The Como argument is problematic conceptually, in particular as concerns its centerpiece, the complementarity of space–time coordination and the claim of causality, and this complementarity was to disappear quickly from Bohr's writings.

It was, however, not complementarity but causality that was a problem. The idea that independent quantum behavior is causal, while the lack of determinism in our predictions concerning quantum phenomena is due to the disturbance brought about by observation, appears to originate, via the transformation theory, with Dirac, who introduced it sometime in 1926 while at Bohr's Institute in Copenhagen (Dirac 1927). Dirac also read the English draft of the Como article, but his input appears to have been minimal, and he also did not appear to have ever warmed up to the concept of complementarity. This view of quantum behavior has been and continues to be pervasive in foundational arguments concerning quantum theory. By contrast, Bohr quickly came to realize the difficulties of sustaining this type of view, not really supported by his argument in the lecture either. Already his next publication abandons the idea and the complementarity of space–time coordination and the claim of causality, thus giving the Como argument barely a yearlong life span. By the time of his reply to EPR, he speaks of "*a final renunciation* of the classical ideal of causality" (Bohr 1935, p. 697; emphasis added) and eventually of complementarity (referring to his overall interpretation of quantum mechanics) as "a rational *generalization* of the … idea of causality" (Bohr 1949, PWNB 2, p. 41; emphasis added). This is quite different from claiming causality at the quantum level, even if only as complementary to the space–time coordination.

First of all, the idealization of both observation and, especially, definition invoked by Bohr is primarily mathematical, and, in the case of definition, it refers to the mathematical formalism of quantum mechanics, such as Schrödinger's equation. While Bohr's statement "*The very nature of the quantum theory thus forces us to regard the space–time co-ordination and the claim of causality, the union of which characterizes the classical theories, as complementary but exclusive features of the description*" introduces the idea of complementarity, Bohr does not rigorously define what he means by "complementary." Specifically, it is not clear what complementary means here beyond "exclusive."

Eventually, Bohr defined complementarity more rigorously in the sense explained from the outset of this study. It designates (a) a mutual exclusivity of certain phenomena, entities, or conceptions; and yet (b) the possibility of considering or applying each one of them separately at any given point; and (c) the necessity of using all of them at different moments for a comprehensive account of the totality of phenomena that we must consider. As I stressed in the Introduction, parts (b) and (c) of this definition are just as important as part (a), and disregarding them often leads to misunderstandings, which are quite common, and the Como argument may have led to some of these misunderstandings. These features of the concept are not stated in the Como lecture, although (c) appears to be implied and to be the reason Bohr sees space–time coordination and the claim of causality not only as "mutually exclusive" but also as "complementary." On the other hand, as will be seen presently, both (b) and (c) pose difficulties in the case of this particular complementarity. However, the definition just given establishes a general concept, which can be instantiated by other complementary quantum configurations, such as that of the (exact) position and momentum measurements or those of "the space–time description and the laws of conservation of energy and momentum" (Bohr 1929, PWNB 1, p. 94). There are two complementarities here, each correlative to the corresponding uncertainty relations. Now all three features—(a), (b), and (c)—rigorously apply.

Now, if the independent behavior of quantum objects is, as Bohr maintains, mutually exclusive with observation, it follows that this behavior is unobservable. How and in what sense, then, could one speak of "the claim of causality" concerning this behavior, at least in physical terms, especially given that one needs something like the classical model to have physical causality? Thus, in classical statistical physics, we can use such a model, established on the basis of the observable behavior of classical object but applied to unobservable constituents of systems containing very large numbers of such constituents. In quantum mechanics, however, we deal with the individual behavior to which, as became clear already with the old quantum theory, the classical model does not apply. One might speak of "mathematical causality," or better, "determination," whereby the equations of quantum mechanics *determine* the relevant *mathematical* object, say, a wave function in the case of Schrödinger's equation (the mathematical part of the "idealization" invoked by Bohr), at any time once it is known at a given time. In this case, one can speak of the mutually exclusive nature of "observation" and "definition," because the formalism itself only refers to the probabilities of the outcomes of what is to be observed (on the basis of previous observation), and never to what is actually observed. Unlike in classical physics, where the same mathematical determination holds as well (and where it also directly relates to observable data), this determination does not translate into a physical causality even when the system in question is undisturbed by measurement, or it least it is not clear how to accomplish such a translation. Bohr, however, appears at least to imply physical causality here, without explaining the requisite mechanism for this translation. In particular, he says that while space–time coordination and the claim of causality are mutually exclusive in quantum physics, the classical theory is characterized by the unity of both. In classical physics, however, the claim of causality is physical, and the equations of classical physics map physically causal processes. Hence, some physical causality, now as mutually exclusive with space–time coordination, appears to be implied by Bohr in quantum physics as well.

What might be ascertained instead is that, unlike in classical physics, in quantum physics the independent behavior (causal or not) of the quantum system and the observation *are mutually exclusive*, in view of the irreducible "disturbance" or, better, "interference" affecting the behavior of quantum systems in any act of observation on which our predictions are based. Even the slightest possible observational "interference"—say, by a single photon—would be sufficient, as Heisenberg explained in his paper on the uncertainty relations (Heisenberg 1927, p. 65). It does, then, follow that this independent behavior of quantum systems is mutually exclusive with observation and hence is unobservable. In classical physics, the independent behavior of a system is not mutually exclusive with observation because the interference in question can, in principle, be neglected or compensated for, as Bohr often stressed. Hence, this behavior can be considered independently and happens to be causal, and the formalism of classical mechanics maps this behavior—at least, again, in the case of idealized models. Bohr is thus right to say that "*the union … of the space–time coordination and the claim of causality … characterizes the classical theories.*" No such models appear to be possible in quantum theory. The situation compelled Bohr to speak of the constitutive, irreducible role of measuring instruments in quantum physics, in contrast to classical physics, a key aspect of all of his interpretations of quantum phenomena and quantum mechanics. But, again, in what sense, then, could one speak of independent quantum behavior as causal? Or, to begin with, in what sense could one meaningfully claim that the formalism of quantum mechanics describes this independent behavior, especially since Bohr insists on the *symbolic* character of this formalism, including Schrödinger's equation, in view of its strictly pre-dictive and, moreover, probabilistically predictive nature?[4]

The problem just discussed is far from inconsequential, and is found elsewhere and is indeed pervasive in the physics community. Bohr's Como argument is not the only source of this view of causality in quantum mechanics. As I said, this view appears to originate with Dirac, who advanced it, via transformation theory, in his influential 1927 paper "The Physical Interpretation of the Quantum Dynamics" (Dirac 1927), completed while at Bohr's institute in Copenhagen in 1926. The paper had a major impact on Heisenberg's thinking and his paper introducing the uncer-tainty relations, where, however, Heisenberg maintained a strong position against quantum level physical causality. Both papers influenced Bohr's thinking at the time, as did Schrödinger's wave mechanics. However, while Schrödinger's program also aimed at a causal theory of quantum processes, his view of the situation was different. The concept of disturbance of quantum processes by observation played no role in his theory, and he hoped that the causal behavior of quantum systems would, at least in principle, be accessible to his wave mechanics. Bohr appears to have been the first to formulate expressly the idea of the juxtaposition, with the corresponding complementarity, of "causality" with "observation" (as "disturbance") in the Como lecture. This juxtaposition was only implied, albeit difficult to miss, in Dirac's paper. However, Dirac adopts Bohr's language and, I would argue, way of thinking in the Como lecture, while not mentioning complementarity (the concept that he, as I said, shuns), in his *The Principles of Quantum Mechanics*, published in 1930 and republished in four subsequent editions, the last in 1958 (Dirac 1958). Thus he says there: "(W)e must revise our ideas about causality. Causality applies only to a [quantum] system which is left undisturbed. If a system is [quantum level] small, we cannot observe it without producing a serious disturbance and hence we cannot expect to find any causal connexion between the results of our observations" (Dirac 1958, p. 4).[5] This is close to Bohr's Como argument and the complementarity of

---

[4] A causal quantum level behavior is a more rigorously established feature in certain alternative interpretations of quantum mechanics, such as the many-worlds interpretation, or alternatives to quantum mechanics itself, in particular Bohmian theories. The latter is consistent with both the underlying causality of the independent quantum behavior and with the view that the probabilistic predictions of the theory, which coincide with those of standard quantum mechanics, arise because of observational disturbance of this behavior, strictly mapped by the formalism (again, at the cost of nonlocality). In the many-worlds interpretation, we observe (as far as our instruments permit us) what actually happens in the way we do in classical physics.

[5] It is not really a matter of the "smallness" of a given system, since it could be large, but only of the "smallness" of its ultimate quantum constituents. Large quantum systems cannot be observed as *quantum systems* without using classically described measuring instruments and hence without these systems being "disturbed," in the way we observe classical systems, by disregarding the role of Planck's constant, *h*. Indeed, classical systems ultimately have a quantum constitution as well, and this constitution is unavailable to classical observation.

space–time coordination (defining observation) and the claim of causality. Among those also adopting Bohr's view and following his language, are Heisenberg in his 1929 Chicago lectures, published in 1930 (Heisenberg 1930), and J. von Neumann in his Mathematical Foundations of Quantum Mechanics (von Neumann 1932). Given their prominence and impact, these works and Dirac's equally influential book might have been especially responsible for the prevalence of this view.

Indeed, even Bohr's argument in the Como lecture is less definitive in this regard than other arguments just mentioned, especially given his view of Schrödinger's wave mechanics as *symbolic* and his discussion of it in the lecture (Bohr 1927, PWNB 1, pp. 73–80). As he says: "In fact, wave mechanics, just as the matrix theory, on this [Bohr's] view represents a *symbolic transcription* of the problem of motion of classical mechanics adapted to the requirements of quantum theory and only to be interpreted by an explicit use of the quantum postulate" (Bohr 1927, PWNB 1, p. 75; emphasis added). From this perspective, Schrödinger's equation or any other part of the mathematical machinery of quantum mechanics cannot be seen as describing or even referring, however indirectly, to the causal behavior of independent quantum systems themselves. As a "mechanics," quantum mechanics, in either form, is only a *symbolic* theory. It assumes the formal structure of classical mechanics, but it only provides a set of algorithms for the predictions, in general probabilistic, concerning the outcomes of certain possible experiments on the basis of previously performed experiments. This formulation is fully in accord with Bohr's subsequent arguments. In the Como lecture, this and several other formulations tend to undermine his argument for the causal character of the independent behavior of quantum objects unless measurement intervenes, which contention, again, grounds most arguments of this type.

This tension further illustrates that, as Bohr was making major steps forward, he was also struggling with many obstacles, which is inevitable in this process, especially, again, given the great complexity of the problems he had to confront. Ultimately, the contributions outweigh the problems; and I would like to close this chapter by reflecting on some of them beyond complementarity itself, the greatest of these contributions.

First of all, the concept of the quantum postulate as *symbolized* by Planck's quantum of action (h)—rather than seeing the quantum of action as reflecting the discontinuous nature of quantum objects—is crucial for the reasons discussed in Sect. 2, most especially in indicating a new form of "atomicity," developed by Bohr into a full-fledged concept later on.

Bohr also brings onto the center-stage the irreducible role of measuring instruments and the radical consequences of this role, in particular in showing that "radiation in free space as well as isolated material particles are abstractions, their properties on the quantum theory being definable and observable only through their interactions with other systems [i.e., classical macrosystems]. Nevertheless,

these abstractions are ... indispensable for a description of experience in connection with our ordinary space–time view" (Bohr 1927, PWNB 1, pp. 56–57).
This formulation still leaves space for attributing such properties to quantum
objects (at the time of measurement), which is short of Bohr's ultimate view, but it
is a radical step, nonetheless, and some of its implications are derived already in
the lecture. Thus, Bohr notes:

> On the whole, it would scarcely seem justifiable, in the case of the interaction problem, to
> demand a visualization by means of ordinary space–time pictures. In fact, all our
> knowledge concerning the internal properties of atoms is derived from experiments on
> their radiation or collision reactions, such that the interpretation of experimental facts
> ultimately depends on the abstraction of radiation in free space, and free material particles.
> Hence, our whole space–time view of physical phenomena, as well as the definition of
> energy and momentum, depends ultimately upon these abstractions. (Bohr 1927, PWNB 1,
> p. 77)

This critical view of particles or wave-like radiation as independent physical
processes is further amplified by his argument, mentioned above, for the symbolic
character of quantum mechanics:

> The symbolic character of Schrödinger's method appears not only from the circumstance
> that its simplicity, similarly to that of the matrix theory, depends essentially on the use of
> imaginary arithmetic[al] quantities. But above all there can be no question of an imme-
> diate connection with our ordinary conceptions because the "geometrical" problem rep-
> resented by the wave equation is associated with the so-called coordinate space, the
> number of dimensions of which is equal to the number of degrees of freedom of the
> system, and, hence, in general greater than the number of dimensions of ordinary space.
> (Bohr 1927, PWNB 1, pp. 76–77)

It is important that Bohr addresses, via the symbolic nature of both theories,
their *physically* and *epistemologically* analogous character, their mathematical
equivalence being already established by that time. This argumentation is not that
far away from his ultimate view, whereby no properties of any kind are attributed
to quantum objects themselves, and hence one no longer would speak of any
knowledge or even conception concerning them, but only of certain effects of their
interactions with measuring instruments upon those instruments. Hence, one only
needs to use some of the classical-like "abstractions" in their application to
measuring instruments, which one can do, since these instruments are described by
means of classical physics.

Bohr's elementary derivation of the uncertainty relations, directly from de
Broglie's formulas, helped to establish them as an experimental fact, a quantitative
law of nature that cannot be circumvented in considering quantum phenomena. An
assumption that it is in principle possible to do so is found in most of Einstein's
criticisms of quantum mechanics, which he, accordingly, saw as incomplete in
view of the fact that the uncertainty relations were embedded in the mathematical
formalism of quantum mechanics. This assumption became one of the main targets
of Bohr's counterarguments. The fact that the uncertainty relations are embedded
in formalism, he argues, only shows that quantum mechanics properly accounts for
this particular aspect of quantum phenomena. While Heisenberg's uncertainty

relations paper also makes this point, Bohr's derivation and analysis of the uncertainty relations in the lecture sharpens it.

Finally, it is worth noting the appearance of certain key terms and concepts that are to remain with Bohr throughout his writings on complementarity, such as "a renunciation as regards the causal space–time coordination of atomic processes," "ambiguity of definition" of variables at the quantum level, or "distinguishing sharply." "Renunciation" becomes a prominent word in Bohr's writings, from his next major contribution, "The Quantum of Action and the Description of Nature" (Bohr 1929), to his reply to EPR (e.g., Bohr 1935, p. 697) and beyond. The argument of the Como lecture that once "in order to make observation possible we permit certain interactions with suitable agencies of measurement, … an unambiguous definition of the state of [a quantum] system is naturally no longer possible" (Bohr 1927, PWNB 1, p. 54) will eventually translate into that against the "essential ambiguity" Bohr locates in the EPR argument.

In sum, although significant refinements and even major changes are introduced as Bohr develops his thinking, the Como lecture establishes the basic framework for this thinking. Most especially, it introduces some of Bohr's key concepts, beginning with complementarity, and prepares the ground for developing these concepts and for the introduction of other concepts, such as phenomenon and atomicity. These concepts also reflect that in his ultimate view of quantum phenomena and quantum mechanics, nothing could any longer be said about the independent behavior of quantum objects, which, again, makes the suspension of causality automatic. Bohr begins to move to this view under the impact of his exchanges with Einstein in 1927. This shift is clearly apparent in the first article published after the Como lecture, "The Quantum of Action and the Description of Nature," discussed in the next chapter. For Bohr, the history of a possible claim of causality in quantum theory had ended and a new history had begun—the history of quantum theory in which any description or conception, let alone a causal one, of quantum objects and their behavior was, as he will say, "*in principle* excluded" (Bohr 1949, PWNB 2, p. 62).

Chapter 5
1929. "The Quantum of Action and the Description of Nature": New Complementarities and a New Interpretation

5.1 Beyond (and Against) the Como Argument

The Como version of complementarity, discussed in the preceding chapter, proved to be short-lived. Within a year or so, even before the Como lecture was published, Bohr was on his way to an essentially different approach, in part shaped by his initial exchanges with Einstein in 1927 (Bohr 1949, *PWNB* 2, pp. 47–52). These exchanges were, I argue here, a decisive factor in changing Bohr's Como argument and in establishing some of the key aspects of Bohr's ultimate view of quantum phenomena and quantum mechanics, which had crystallized, under the impact of his subsequent exchanges with Einstein, by the late 1930s. Bohr's new argumentation was first presented in print in his 1929 article "The Quantum of Action and the Description of Nature," originally written for the issue of *Die Naturwissenschaften* in celebration of the fiftieth doctoral anniversary of Max Planck. The article, to which this chapter is primarily devoted, was Bohr's first published work on complementarity after his exchanges with Einstein just mentioned and it was clearly shaped by them, as Bohr noted in his later account of these exchanges (Bohr 1949, *PWNB* 2, p. 52). The double-slit experiment that played a major role in these exchanges did not, however, figure in this article, and only assumed a major role in Bohr's subsequent communications, beginning with his important but never published Bristol lecture of 1931 "Space–Time Continuity and Atomic Physics" (Bohr 1972–1999, v. 6, pp. 361–370). For this reason, the double-slit experiment will be discussed in the next chapter, which will also consider Bohr's 1927 exchanges with Einstein, as presented by Bohr in his 1949 "Discussion with Einstein." In any event, "The Quantum of Action and the Description of Nature" represents a crucial step in the development in Bohr's thinking.

Bohr's argument in the article is based on a new form of complementarity, a new instantiation of the concept—the complementarity of "the space–time description and the laws of conservation of energy and momentum," in correspondence with, and as an interpretation of, Heisenberg's uncertainty relations (Bohr 1929a, *PWNB* 1, p. 94). This complementarity is not found in the Como

A. Plotnitsky, *Niels Bohr and Complementarity*, SpringerBriefs in Physics, DOI: 10.1007/978-1-4614-4517-3_5, © Arkady Plotnitsky 2013

lecture, based, as discussed in Chap. 2, on the complementarity of the space–time coordination and the claim of causality, which, I argue, is problematic and even unsustainable, as Bohr came to realize sometime in 1927–1928. "The Quantum of Action and the Description of Nature" abandons this complementarity and, to begin with, the idea of the independent causal behavior of quantum objects. Bohr's new complementary architecture is defined by different, mutually exclusive, experimental arrangements. It is also essentially linked to the irreducibly probabilistic character, established *on experimental grounds*, of our predictions concerning quantum phenomena, even and in particular those related to individual quantum events, a character that is in turn correlative to the uncertainty relations. By the same token, quantum processes themselves are now seen as noncausal in all circumstances, and by the time Bohr arrives at his ultimate view of quantum phenomena in the late 1930s, quantum objects and their behavior are placed beyond any possible description or even conception.

The interpretation of quantum phenomena and quantum mechanics offered in "The Quantum of Action and the Description of Nature" may be seen as an intermediate or transitional one between the Como version and Bohr's ultimate version of his interpretation. This version is, however, much closer to Bohr's ultimate version than it is to the Como version, although, as I said, it still contains echoes of the Como argument. Thus, Bohr says that "any attempt at ordering in space–time leads to a break of the *causal* chain" (Bohr 1929a, *PWNB* 1, p. 98). Bohr's surrounding elaboration, however, appears to leave space for reading the statement in a conditional sense, to the effect that such would be the case were one to assume that quantum processes are causal (Bohr 1929a, *PWNB* 1, pp. 97–98). I shall not insist on either reading, because it is possible that Bohr's earlier ideas were still affecting his thinking, and moreover, the particular elaboration where this statement occurs relates to a possible use of complementarity in psychology, where the Como-type argument may in fact be applicable. As will be seen in Chap. 9, however, even in the case of psychology, Bohr ultimately shifts to different instantiations of complementarity, which are much closer to the approach initiated by the article under discussion (Bohr 1954, *PWNB* 2, pp. 77–79). In any event, it is hardly in doubt that the article significantly departs from the Como argument and reflects a very different way of thinking concerning quantum phenomena and quantum mechanics.

Bohr changed the term "complementarity" to "reciprocity" in the article. This change appears to be motivated by the problems of the Como lecture and by the kind of response it received in the physics community. Apart from those close to Bohr, such as Heisenberg and Pauli, and a few others, such as Paul Ehrenfest, the reception of Bohr's article was relatively cold. The lecture did not appear to most to introduce a new physics. This, as I argued in Chap. 4, is not true, although it is true that the new physics of Bohr's argument was difficult to appreciate or even to perceive without properly following the philosophical considerations involved in his argument. Nor did the lecture offer new mathematical tools that would help one to deal with the still outstanding problems of quantum mechanics, or a new version of the mathematical formalism of quantum mechanics. This assessment was correct, although Bohr did not aim at doing so. The difficulty of Bohr's exposition was a factor as well. It took a while and a few subsequent interventions by Bohr before the deeper philosophical ideas and implications of his argument took hold. Many, on both sides (pro and contra Bohr's argument), were also troubled by the radical implications of the argument. These worries have never subsided, while Bohr's epistemology only became more radical as his argumentation developed. For the moment, the term "reciprocity" appears to be indebted to the uncertainty relations, which entail "reciprocal" (mutually influencing) relations between the quantities involved. Bohr, who came to see the change as a mistake, quickly returned to the language of complementarity, specifically in his 1931 Bristol lecture, "Space–Time Continuity and Atomic Physics."

Bohr's new argument, especially as coupled to his analysis of the double-slit experiment, as it is in the Bristol lecture, also gave Bohr a platform for his response to EPR a few years after. In referring to his earlier publications in his reply to EPR, Bohr says: "I shall therefore be glad to use this opportunity to explain in somewhat greater detail a general viewpoint, conveniently termed 'complementarity,' which I have indicated on various previous occasions" (Bohr 1935, p. 696, n.). Given Bohr's actual argument in his reply, this statement, accompanied by a reference to Bohr's 1934 collection, *Atomic Theory and the Description of Nature* (*PWNB* 1), requires qualification. Thus, it would be especially difficult to apply this statement to the Como lecture. It can, however, *nearly* apply to "The Quantum of Action and the Description of Nature," contained in that volume or to Bohr's "Introductory Survey" (Bohr 1929b) there and the final article in the volume, which also dates from 1929, or, again, to the 1931 Bristol lecture. I qualify because his reply to EPR introduces yet further changes into Bohr's argumentation and thus makes further steps toward his ultimate understanding of the quantum-mechanical situation.

The shift in Bohr's views by 1929 and his partial return to the position he took in response to Heisenberg's discovery of quantum mechanics in "Atomic Theory and Mechanics" (1925), also included in the volume, are clearly confirmed by the articles just mentioned. Thus, Bohr's "Introductory Survey" (Bohr 1929b) not only adopts his approach developed in "The Quantum of Action and the Description of Nature," but also recasts the Como lecture in these new terms by stressing those aspects of its argument that were retained in his new approach. Most notably, the complementarity of space–time coordination and the claim of causality, central to the Como lecture, is never mentioned. Bohr speaks instead of our being forced "step by step to forego a causal description of the behavior of individual atoms in space and time" and "the renunciation of causality in the quantum-mechanical description" (Bohr 1929b, *PWNB* 1, p. 4).

5.2 A Renunciation of the Causal Space–Time Mode of Description

I would now like to discuss the setup and the main gradient of "The Quantum of Action and the Description of Nature," as against the Como lecture. First, the language of "the quantum postulate" is supplemented by that of "the elementary quantum of action," to be used from now on in Bohr's work (Bohr 1929a, *PWNB* 1, p. 92). In his "Introductory Survey," Bohr notes:

> [T]he application of [classical mechanics and electromagnetic theory] to atomic problems was destined to reveal a hitherto unnoticed limitation that found its expression in Planck's discovery of the so-called quantum of action, which imposes upon individual atomic processes an element of discontinuity quite foreign to the fundamental principles of classical physics, according to which all action may vary in a continuous manner. The quantum of action has become increasingly indispensable in the ordering of our experimental knowledge of the properties of atoms. (Bohr 1929b, *PWNB* 1, p. 4)

From the outset, Bohr (this continues the Como argument) defines quantum mechanics as a "symbolic" theory, with which the history of quantum theory "has reached a temporary climax." The climax is seen as temporary because of the remaining problems of interpretation and because of quantum electrodynamics, which was introduced by that time, and posed new problems, albeit not affecting quantum mechanics. Although radically different from classical mechanics in its key physical, mathematical, and epistemological features, quantum mechanics is argued to be "a natural generalization of the classical mechanics with which in beauty and self-consistency it may well be compared" (Bohr 1929a, *PWNB* 1, p. 92). Bohr then places quantum theory within the history of our attempts (such as in the kinetic theory of gases), "to accomplish [a causal space–time] description also in the case of phenomena, which, in our immediate sense impressions, do not appear as motions of material bodies," in contrast to phenomena considered in classical mechanics (Bohr 1929a, *PWNB* 1, pp. 92–93). The question becomes whether it is possible in the case of quantum phenomena to make the type of assumption concerning the causal behavior of primitive individual elements that we can make in classical statistical physics or whether we must renounce this type of assumption. Bohr says:

> This goal has not been attained, still, without a renunciation of the causal space–time mode of description that characterizes the classical physical theories which have experienced such a profound clarification through the theory of relativity. In this respect, the quantum theory may be said to be a disappointment, for the atomic theory arose just from the attempt to accomplish such a description also in the case of phenomena which, in our immediate sense impressions, do not appear as motions of material bodies. From the very beginning, however, one was not unprepared in this domain to come upon a failure of the forms of perception adapted to our ordinary sense impressions. We know now, it is true, that the often expressed skepticism with regard to the reality of atoms was exaggerated; for, indeed, the wonderful development of the art of experimentation has enabled us to study the effects of individual atoms. Nevertheless, the very recognition of the limited divisibility of physical processes, symbolized by the quantum of action, has justified the

old doubt as to the range of our ordinary forms of perception when applied to atomic phenomena. Since, in the observation of these phenomena, we cannot neglect the interaction between the object and the instrument of observation, the question of the possibilities of observation again comes to the foreground. Thus, we meet here, in a new light, the problem of the objectivity of phenomena which has always attracted so much attention in philosophical discussion. (Bohr 1929a, *PWNB* 1, pp. 92–93)

The contrast with the Como argument is striking. The complementarity of space–time coordination and the claim of causality is gone, along with causality itself. What replaces them, as the grounding feature of quantum epistemology, is the crucial *distinction*, which is to shape Bohr's argumentation from this point on, between unobservable quantum *objects* and observable *phenomena*, defined through the *effects* of the interactions between quantum objects and measuring instruments, observed in these instruments. It is in principle possible, under these conditions, to speak of certain properties of quantum objects, unobservable but ascertainable through these effects at least at the time of measurement, which view Bohr eventually abandons as well, following the EPR argument. One can especially note the language of *effects*, which was to become decisive for Bohr's later works. If "the wonderful development of the art of experimentation has enabled us to study the effects of individual atoms," it has also limited us to the observation or descriptive study of these effects (predicted by means of the quantum theory). As quantum objects, atoms themselves cannot be studied or even rigorously described or, in Bohr's ultimate view, conceived of as quantum objects. Nor, for that matter, can be any entity, considered as a quantum object, which, again, can be macroscopic in scale, although its quantum nature is defined by their ultimate microscopic constitution. This nature may be more pointedly manifest via certain effects, as in the case of Josephson's junctures, or not, as in the case of more conventional macro objects, from everyday objects to planets to stars to galaxies to the Universe itself. All these entities, beginning with atoms, exist as quantum objects. However, in this interpretation, if such entities can be observed, they can only be observed as *classical* objects, while, as quantum objects, they remain unobservable and indescribable, or even inconceivable. Their quantum nature could only manifest itself in certain effects these objects produce in measuring instruments by interacting with them.

While classical physics and even (with important qualifications) relativity may be seen as a refinement of our everyday experience and language, or of "the forms of perception adapted to our ordinary sense impressions," such forms of perception, however refined, are not applicable to quantum objects. (Bohr notes an epistemological more radical character of quantum theory vis-à-vis relativity later in the article [Bohr 1929a, *PWNB* 1, pp. 97–98]). The argument is amplified in Bohr's commentary, cited earlier, introducing the article in "Introductory Survey," where he argues "that all our ordinary verbal expressions bear the stamp of our customary forms of perception, from the point of view of which the existence of the quantum of action is an irrationality" (Bohr 1929b, *PWNB* 1, p. 19). As noted earlier, Heisenberg makes a similar point in his 1929 Chicago lectures (Heisenberg 1930, p. 11). "Discussion with Einstein" gives Bohr's arguably strongest expression of this idea: "the peculiar individuality of quantum effects presents us, as regards the comprehension of well-defined evidence, with a novel situation unforeseen in classical physics and irreconcilable with conventional ideas suited for our orientation and adjustment to ordinary experience [on which classical physics is based]. It is in this respect that quantum theory has called for a renewed revision of the foundation for the unambiguous use of elementary concepts as a further step in the development which, since the advent of relativity theory, has been so characteristic of modern science" (Bohr 1949, *PWNB* 2, p. 62).

The final sentence of Bohr's passage from "The Quantum of Action and the Description of Nature" cited above ("Thus, we meet here, in a new light, the problem of the objectivity of phenomena which has always attracted so much attention in philosophical discussion") refers to Kant's epistemology of phenomena (or appearances) and noumena (or things-in-themselves), but also indicates a possible move to a more radical epistemology. While unknowable, Kant's things-in-themselves are still thinkable, which also implies that, in principle, our thinking could correspond to the nature of reality, even if we cannot be completely assured that it does (Kant 1995, p. 115). As discussed earlier, Einstein essentially takes the same view. Bohr now moves toward a more radical view, crystallized in his later works, whereby quantum objects and their behavior are places beyond the reach not only of knowledge but also of thought itself. This epistemology is also Bohr's ultimate response to the great difficulties and, ultimately, the impossibility of applying classical-like visualizable [*anschaulich*] mechanical pictures in quantum physics. The protracted history of dealing with this situation is one of the main themes of the articles assembled in *Atomic Theory and the Description of Nature*, beginning with "Atomic Theory and Mechanics" (1925). "The Quantum of Action and the Description of Nature" contains important elaborations dealing with the question of visualization [*Anschaulichkeit*] or, as the case may be, the unvisualizability of quantum processes, "the resignation with regard to the desires for visualization," as Bohr says (Bohr 1929a, *PWNB* 1, p. 98). Bohr retains "objectivity" at the level of effects of the interactions between quantum objects and measuring instruments, effects that are defined classically and, thus, ensure "objectivity" as the possibility of "unambiguous communication" (Bohr 1954, *PWNB* 2, p. 67). I shall return to this point, more pronounced in Bohr's later articles, in Chap. 10.

For the moment, this type of epistemology, Bohr argues in "The Quantum of Action and the Description of Nature," entails, correlatively, both the irreducibly probabilistic nature of quantum mechanics and the irreducible complementary nature of certain features of description. Now, however, as against the Como lecture, these features defined in terms of certain experimentally defined phenomena, in a strict correspondence with the uncertainty relations, the main quantitative law of quantum mechanics (Bohr 1929a, *PWNB* 1, p. 95). According to Bohr (who, again, replaces "complementarity" with "reciprocity"):

[W]e know now that for material particles as well as for light different conceptual pictures are necessary to account completely for the phenomena and to furnish a unique formulation of the statistical laws which govern the nature of observation. The more clearly it appears that a uniform formulation of the quantum theory in classical terms is impossible, the more we admire Planck's happy intuition in coining the term "quantum of action" which directly indicates a renunciation of the [continuous] action principle, the central position of which in the classical description of nature he himself emphasized on more than one occasion. This principle symbolizes, as it were, the peculiar reciprocal symmetry relation between the space–time description and the laws of the conservation of energy and momentum, the great fruitfulness of which, already in classical physics, depends upon the fact that one may extensively apply them without following the course of phenomena in space and time. It is this very reciprocity [complementarity] which has been made use of in a most pregnant way in the quantum–mechanical formalism. As a matter of fact, the quantum of action [h] appears here only in relations in which space–time coordinates and momentum-energy components, which are canonically conjugate quantities in the Hamiltonian sense, enter in a symmetrical and reciprocal [complementary] manner. (Bohr 1929a, *PWNB* 1, p. 94)

Bohr, thus, clearly shifts to a new form of complementary configurations, defined by kinematical variables, on the one hand, and conservation laws, on the other, as against the complementarity of the space–time coordination and the claim of causality, dominant in the Como lecture. Bohr is of course right to say, that "the great fruitfulness of [the laws of conservation of energy and momentum], already in classical physics, depends upon the fact that one may extensively apply them without following the course of phenomena in space and time." The point, however, is that in quantum theory, where these laws apply with equally great fruitfulness, one not only *may* apply them without following the course of phenomena, or in this case, objects, in space and time, but also *must* apply them without doing so, at least in Bohr's interpretation. The argument is developed in his Bristol lecture, "Space–Time Continuity and Atomic Physics," helpfully accompanied by Bohr's return to the language of complementarity: "We have thus either space–time description or description where we can use the laws of conservation of energy and momentum. [These descriptions] are *complementary* to each other" (Bohr 1931, p. 369).

5.3 Classical Concepts, the Uncertainty Relations, and Probability

Bohr moves next to the role of classical (physical) concepts and the uncertainty relations, as correlative to the complementarity of the space–time description and dynamical conservation laws. He notes first: "It lies in the nature of physical observation, nevertheless, that all experience must ultimately be expressed in terms of classical concepts, neglecting the quantum of action. It is, therefore, an inevitable consequence of the limited applicability of the classical concepts that the results attainable by any measurement of atomic quantities are subject to an inherent limitation" (Bohr 1929a, *PWNB* 1, pp. 94–95). This is an important formulation, especially given that Bohr's appeal to classical concepts in this context is often misunderstood. I shall discuss the subject later, but a few key points are worth summarizing in the context of the article under discussion, in order to present the import of Bohr's argument more fully.

First, this formulation clearly states the following point, often missed by commentators on Bohr. Although indispensable, classical concepts are never sufficient for a proper quantum-mechanical account, as is shown by Heisenberg's uncertainty relations to which Bohr proceeds from this point in the article itself. Second, and by the same token (this point is also usually missed by commentators), in Bohr's ultimate view, quantum objects or their quantum interactions with measuring instruments are never subject to description in terms of classical physical or any other concepts. Bohr's earlier view, including in the article under discussion, allows an attribution of certain properties to quantum objects at the time of measurements (under the constraints of the uncertainty relations), but not independently. In his ultimate view, any such description can only apply to these measuring instruments, onto which the physical application of the uncertainty relations is now transferred, or, again, more accurately, to a certain classically described stratum of the instruments, since these instruments are seen as having a quantum stratum through which they interact with quantum objects. This interaction is "irreversibly amplified" to the classical level, say, as manifest by a spot left on a silver screen (e.g., Bohr 1954, *PWNB* 2, p. 73). Finally, classical physics is seen as a refinement of our everyday perception and thinking, which, effective as they are in classical physics or, to a more limited extent, in relativity, may not be suitable for the quantum scale of nature, no matter how far this refinement may reach. According to Bohr, the uncertainty relations reflect this situation:

> A profound clarification of this question was recently accomplished with the help of the general quantum-mechanical law, formulated by Heisenberg, according to which the product of the mean errors with which two canonically conjugate mechanical quantities may be simultaneously measured can never be smaller than the quantum of action. Heisenberg has rightly compared the significance of this law of reciprocal uncertainty for estimating the self-consistency of quantum mechanics with the significance of the impossibility of transmitting signals with a velocity greater than that of light for testing the self-consistency of the theory of relativity. In considering the well-known paradoxes which are encountered in the application of the quantum theory to atomic structure, it is essential to remember, in this connection, that the properties of atoms are always obtained by observing their reactions under collisions or under the influence of radiation, and that the above-mentioned limitation on the possibilities of measurement is directly related to the apparent contradictions which have been revealed in the discussion of the nature of light and of material particles. (Bohr 1929, *PWNB* 1, p. 95)

These contradictions are, however, only apparent, or more accurately, quantum mechanics allows us to offer a non-contradictory theory of the data that this situation defines, even though, if one speaks of it as a *physical* theory, this may only be possible if the formalism is accompanied by a suitable interpretation, which also contains an interpretation of quantum phenomena themselves. In that case, the interpretation is based on the impossibility of assigning customary physical properties to quantum objects apart from measurement and, in Bohr's ultimate interpretation, even then (all physical properties being considered as those of measuring instruments themselves). Bohr's argumentation, indicated in this passage, takes significant steps, especially as against the Como argument, in resolving the contradictions in question, although it does not quite reach the level of his post-EPR works, where, again, assigning such properties is abandoned altogether, even at the time of measurement. Heisenberg's γ-ray microscope thought-experiment, alluded to here, is part of the difficulty, even though Bohr had straightened out some of the problems of Heisenberg's initial analysis of this experiment earlier. Heisenberg's thought experiment is helpful insofar as it reflects the fact that the role of the agencies of observation cannot be neglected and, thus, *irreducibly* shape any observable phenomena in quantum physics, and that, while, as observed phenomena, all such phenomena are *classical*, they are the effects of the *quantum* interaction between quantum objects and measuring instruments. However, the experiment may also be misleading insofar as it suggests that one could speak of the "undisturbed" independent behavior of quantum objects, especially as described by the formalism of quantum mechanics, before these objects are "disturbed" by an experiment.

Bohr inches toward his later views by noting that any observation in quantum theory is essentially linked to a *reaction* of quantum objects upon other quantum objects, objects eventually to be seen by Bohr as the quantum aspects of measuring instruments. This transition is also suggested by the statement that builds on the passage just cited, and refines a similar point in the Como lecture. Bohr says: "At the conclusion of[the Como lecture], it was pointed out that a close connection exists between the failure of our forms of perception, which is founded on the impossibility of a strict separation of phenomena and means of observation, and the general limits of man's capacity to create concepts, which have their root in our differentiation between subject and object" (Bohr 1929a, *PWNB* 1, pp. 95–96; Bohr 1927, *PWNB* 1, pp. 90–91). This statement, or that closing the Como lecture, has a broader philosophical import, which I shall discuss in Chap. 9. My main point at the moment is that, along with and in part through our perceptual and conceptual limitations, these elaborations intimate Bohr's later concept of the wholeness or indivisibility of phenomena—the impossibility of "extracting" the independent identity of a quantum object or behavior from what is observed in measuring instruments. On the other hand, the statement "the properties of atoms are always obtained by observing their reactions under collisions or under the influence of radiation" stops short of his later view of the situation via the concept of phenomena. This statement suggests that quantum measurements or even (by way of probabilistic predictions) quantum-mechanical formalism refers to the properties of quantum objects at the time of measurement, as opposed to, as in his later view, the effects of the interactions between quantum objects and measuring instruments, effects manifest *only* in these instruments.

Bohr makes further steps in the direction of his ultimate view in both articles, and in "The Atomic Theory and the Fundamental Principles Underlying the Description of Nature," the final essay of *The Atomic Theory and the Description of Nature*, and the Bristol lecture "Space–Time Continuity and Atomic Physics." In these articles he returns to the language of complementarity. The language of reciprocity was clearly not helpful, as Bohr quickly came to realize.

"The Quantum of Action and the Description of Nature" is, however, especially decisive for the development of Bohr's thinking in this new direction. That "Introductory Survey" devotes the largest space to this article is indicative of its significance (Bohr 1929b, *PWNB* 1, pp. 15–21). Bohr offers there several important elaborations concerning the relationships among quantum–mechanical formalism, classical concepts, probability, the uncertainty relations, complementarity, the role of measuring instruments, and of the distinction between them and quantum objects. Thus he says:

> [Quantum] phenomena belong, indeed, to a domain in which it is essential to take into account the quantum of action and where an unambiguous description [of quantum objects and processes] is impossible. ... [O]nly with the help of classical ideas is it possible to ascribe an unambiguous meaning to the results of observations. We shall, therefore, always be concerned with applying probability considerations to the outcome of experiments which may be interpreted in terms of such conceptions. Consequently, the use made of the symbolic expedients will in each individual case depend upon the particular circumstances pertaining to the experimental arrangement. Now, what gives to the quantum-theoretical description its peculiar characteristic is just this, that in order to evade the quantum of action we must use separate experimental arrangements to obtain accurate measurements of the different quantities, the simultaneous knowledge of which would be required for a complete description based upon the classical theories, and, further, that these experimental results cannot be supplemented by repeated measurement. In fact, the indivisibility of the quantum of action demands that, when any individual result of measurements is interpreted in terms of classical conceptions, a certain amount of latitude be allowed in our account of the mutual action between the objects and the means of observation. This implies that a subsequent measurement to a certain degree deprives the information given by a previous measurement of its significance for predicting the future course of the phenomena. Obviously, these facts not only set a limit to the *extent* of the information obtainable by measurements, but they also set a limit to the *meaning* which we may attribute to such information. We meet here in a new light the old truth that in our description of nature the purpose is not to disclose the real essence of phenomena [i.e., the quantum character of their ultimate constitution] but only to track down, so far as it is possible, relations between the manifold aspects of our experience. (Bohr 1929b, *PWNB* 1, pp. 17–18)

One might of course argue that one could, in principle, go further into "the real essence of phenomena." Indeed, quantum mechanics and Bohr's interpretation of it have done so as well. However, quantum mechanics appears to and, in Bohr's interpretation does, establish certain irreducible limitations upon our capacity to reach the ultimate quantum constitution of nature. In this regard, this elaboration is a bridge to Bohr's ultimate perspective, which will further qualify this argumentation, along the lines just suggested. Bohr's statement that "a subsequent measurement to a certain degree deprives the information given by a previous measurement of its significance for predicting the future course of the phenomena" reflects another crucial aspect of quantum theory, which becomes especially significant in the context of the EPR experiment as well. Any act of measurement discontinuously resets the future evolution of the system. The process starts anew with each new measurement, which erases the outcome of the previous measurements as meaningful for future predictions concerning the system.

It may not be altogether accurate to speak here of "the future *course* of the phenomena." The word "course" implies some form of continuity. By contrast, in this case, one deals with future phenomena (always individual and discrete) defined by possible measurements, the outcomes of which we can predict, but only in probabilistic terms, regardless of what theory we use to make such predictions. In "The Quantum of Action and the Description of Nature" this fact is *interpreted*, as it will be from this point on by Bohr, as a consequence of the irreducible role of measuring instruments in the constitution of quantum phenomena. In the passage cited earlier, Bohr argues that "in the observation of these [quantum-mechanical] phenomena, we cannot neglect the interaction between the objects and the instruments of observation," and that, "this being the state of affairs, it is not surprising that, in all rational applications of the quantum theory, we have been concerned with essentially statistical problems. Indeed, in the original researches of Planck, it was, above all, the necessity of modifying the classical statistical mechanics which gave rise to the introduction of the quantum of action" (Bohr 1929a, *PWNB* 1, p. 93). In Bohr's view, this modification reflects the fact that, unlike in classical statistical mechanics, "the recourse to probability laws under such circumstances is essentially different in aim from the familiar application of statistical considerations as practical means of accounting for the properties of mechanical systems of great structural complexity." Instead, Bohr argues, "in quantum physics we are presented not with intricacies of this kind, but with the inability of the classical frame of concepts to comprise the peculiar feature of indivisibility, or 'individuality,' characterizing the elementary processes" (Bohr 1949, *PWNB* 2, p. 34).

This individuality is, according to Bohr, *symbolized* by Planck's quantum of action. In Bohr's interpretation, this situation, which is, again, correlative to the irreducible role of the measuring instruments in the constitution of quantum phenomena, no longer allows for a causal or any other explanation of the nature and behavior of quantum objects. As stressed from the outset of this study, it follows that we are limited to probabilistic predictions even as concerns primitive individual quantum phenomena and events, in other words, to exactly as much as quantum mechanics provides. Bohr's position in his debate with Einstein was defined by his argument that nature *might* (no stronger claim is possible here) not allow us to do better than this in dealing with quantum phenomena, precisely by *precluding* a description of the processes ultimately responsible for these phenomena (Bohr 1949, *PWNB* 2, p. 62). This possibly uncircumventable limit is not only in accordance with Bohr's interpretation of the situation, but is also the main reason for this interpretation, in the form it takes (as against the Como version) from "The Quantum of Action and the Description of Nature" on. In this view, quantum mechanics would, within its proper scope, be a complete theory of quantum phenomena, as complete as nature allows such a theory to be. This, however, was not a view that Einstein was ever willing to accept. The absence of a more complete theory, by his classical-like criteria, never deterred him any more than this absence (we still have no such theory) deters his followers now.

Chapter 6
1931. "The Space–Time Continuity and Atomic Physics" (the Bristol Lecture): Quantum Phenomena and the Double-Slit Experiment

6.1 Bohr's Epistemology and the Double-Slit Experiment

Bohr's ultimate understanding of quantum phenomena and quantum mechanics gradually emerged, I argue here, under the impact of his exchanges with Einstein, beginning with those that took place in 1927 following the Como lecture and continuing into the 1940s. The double-slit experiment enters these exchanges at the outset never to leave them or Bohr's thinking itself. The experiment did not figure in the Como lecture, or in the preceding work on quantum mechanics by Heisenberg, Schrödinger, and others, on which the Como argument was based. Nor was it used in "The Quantum of Action and the Description of Nature," discussed in the preceding chapter. It was, however, used in the Bristol lecture "The Space–Time Continuity and Atomic Physics" of 1931, which may be seen as the first sustained argument by Bohr to use it. The main reason for Bohr's persistent appeal to the experiment is that it can be effectively used to test our claims concerning quantum phenomena and quantum mechanics, which properly predicts the numerical data found in the double-slit experiment and thus responds to the peculiar character of the phenomena observed there. Once a given argument concerning either quantum phenomena or quantum mechanics leads to a conflict with these features, this argument may be set aside as something that is in conflict with the experimental evidence.[1] In particular, any attempt to circumvent Heisenberg's uncertainty relations, $\Delta q \Delta p \cong h$, leads to this type of inconsistency with the double-slit experiment. The physical meaning of the uncertainty relations

---

[1] While the double-slit experiment was not actually performed as a quantum experiment until later (in the case of electrons in 1960s), other quantum experiments that had been performed could be considered as equivalent to it with respect to the key features of quantum phenomena at stake, which enabled one to use the double-slit experiment as a thought experiment. Cf., Bohr's comments in his reply to EPR (Bohr 1935, p. 698, n.). This is not unusual. That the EPR experiment, as originally proposed by EPR, cannot in principle be performed in a laboratory has never put in doubt its legitimacy for the theoretical arguments based on it. Related experiments, based on Bohm's version of the EPR experiment for spin have been performed, as were experiments statistically approximating the EPR experiment.

A. Plotnitsky, *Niels Bohr and Complementarity*, SpringerBriefs in Physics, 71
DOI: 10.1007/978-1-4614-4517-3_6, © Arkady Plotnitsky 2013

is a subtle matter, which I shall address below. The main point here is that Bohr saw the uncertainty relations as experimentally given, and the fact that they can be derived from quantum mechanics as further testimony that the theory adequately reflects the actual character of quantum phenomena. The uncertainty relations and the data observed in the double-slit experiment are equivalent, a fact used by Bohr throughout his arguments (e.g., Bohr 1935, pp. 697–700; Bohr 1949, *PWNB* 2, pp. 43–47, 52–61).

The double-slit experiment is considered to be a paradigmatic—or, it is some-times argued, even *the* paradigmatic—quantum experiment, in which the famously strange features of quantum phenomena manifest themselves. It is, accordingly, not surprising that the experiment, along with these features and quantum mechanics, lends itself to different interpretations, and it was interpreted somewhat differently by Bohr himself, following the evolution of his views, as considered in this study. Bohr's first major work that expressly bases its argument on the double-slit exper-iment is, again, Bohr's Bristol lecture given in 1931, which was not published during his life-time but the draft of which is available in *Collected Works of Niels Bohr* (Bohr 1972–1999, v. 6, pp. 361–370). It also represents the first sustained example of Bohr bringing together the double-slit experiment and his post-Como thinking, a coupling that continued to define his argumentation from this point on. This is why the lecture and its date figure in my title here.

It is true (and is one of my points) that Bohr's use of the double-slit experiment goes back to his earlier exchanges with Einstein in the 1927 Solvay Conference in Brussels. However, it does not appear that his post-Como views were quite in place in these exchanges. On the other hand, it could be ascertained that these views were in place by 1929. In addition, while there are more contemporaneous accounts, for example, intriguing notes by Paul Ehrenfest (Bohr 1972–1999, v. 6, pp. 37–41), partial transcripts of the discussion of the Solvay Conference (Bohr 1972–1999, v. 6, pp. 100–106), and other records in the *Niels Bohr Archive*, they are fragmentary. Although these accounts have been discussed by commentators, it is difficult to use them for offering a cohesive account of the double-slit experiment from Bohr's perspective. Indeed, it does not appear that Bohr had such an account at the time, except possibly as developed in the course of the discussion that took place there. The Bristol lecture offers a more comprehensive account of the double-slit experiment, arguably close to these exchanges, judging by Bohr's own most sustained account in his 1949 "Discussions with Einstein," which, however, is inflected by his ultimate views, fully in place by that time. My discussion of the double-slit experiment in this chapter also presents it in the way Bohr would ultimately see it in "Discussion with Einstein," after his nearly life-long debate with Einstein, in which Bohr relied on the experiment. This is advantageous, given that Bohr's ultimate epistemology leads to his most consistent view of the experiment. Nor does this approach pose significant historical difficulties because this account could be easily adjusted to correspond to his earlier available accounts of the double-slit experiment without losing the most essential points, which are in place in all of these accounts. Indeed, I would contend that Bohr's ultimate view of the double-slit experiment helps to bring out these points more effectively. I shall, of course, indicate those aspects of Bohr's earlier views that are germane for understanding Bohr's earlier uses of the experiment.

All of Bohr's discussions of the double-slit experiment, even the more technical ones, always proceed by using concepts and language of daily life or those of classical physics (which, Bohr argues, refine those of daily life)—*as far as it is possible to do so.* This qualification is crucial and, I argue here, applies to Bohr's understanding of quantum phenomena in general, often misunderstood on account of Bohr's persistent appeal to both daily and classical-physical language and concepts in quantum theory. The point here is not that our non-technical accounts of quantum phenomena and quantum mechanics are, unavoidably, limited as concerns their explanatory capacity, which is of course true. Nor is the point only that his use of the concepts and language of daily life and classical physics is unavoidable even in technical discussions, although this point is important to Bohr's arguments. Instead the point is that this language and these concepts appear to be ultimately inapplicable to quantum *objects*, which are responsible for quantum *phenomena*, observed in measuring instruments impacted by quantum objects. Contrary to the persistent view, there is no contradiction between the two claims just made.[2] We have no other language and concepts, as far as our language and concepts reflect what can, in principle, be thought. Hence, we must use them in our arguments for the existence of the indescribable and even inconceivable quantum objects on the basis of a particular character of certain phenomena observed in measuring instruments under the impact of these objects. Indeed, we can do not more than infer the existence of such objects on the basis of these effects and in order to explain them. It is in this way that the unthinkable becomes part of what can be thought. For the sake of convenience and economy, I shall sometimes also provisionally use, as Bohr did, classical language and concepts in describing the behavior of quantum objects, when, for example, speaking of "photons passing through slits," while assuming that this language or concept ultimately does not apply and, whenever necessary, explaining what such expressions actually refer to in quantum situations.

6.2 The Double-Slit Experiment: Physics and Epistemology

The double-slit experiment can be performed with all quantum objects (even composite ones, such as carbon 60 fullerene molecules), although it was not actually performed as a quantum experiment with anything other than light until the 1960s. Before then, it had, as I noted, functioned as a thought experiment, without much doubt that it could in principle be performed on any type of quantum object. This confidence was further supported by other key quantum experiments,

---

[2] Heisenberg addressed this subject, which he saw as important in clearing up misconceptions about quantum mechanics, in his response to the EPR argument (Heisenberg 1935).

such as the Stern-Gerlach experiment, various experiments in quantum interferometry, the beam-splitter experiment and other experiments with half-silvered mirrors, which had been performed earlier and which exhibit the main features of quantum phenomena exhibited in the double-slit experiment. One of the advantages of the double-slit experiment is that it can be especially easily explained *qualitatively* without technical knowledge of quantum theory. Properly predicting the corresponding *quantitative* data would, of course, require some theory, such as quantum mechanics, or quantum electrodynamics in the case of photons, which cannot be treated by means of quantum mechanics.[3]

The experimental arrangement of the double-slit experiment consists of a source, such as that of a monochromatic light (which makes it possible to emit photons one by one), and, at some distance from it, a diaphragm with a single slit (A); at a sufficient distance from it a diaphragm with two slits (B and C), widely separated; and finally, at a sufficient distance from the diaphragm, a screen, a silver bromide photographic plate (Fig. 6.1). Technically, one does not need the first diaphragm to illustrate the key features of the experiment and can use the source itself to define the initial stage of the experiment. However, the arrangement described here is convenient if one wants to relate the experiment to the uncertainty relations, and in part for that reason, it is used by Bohr in most of his arguments. Two setups are considered, in each of which a sufficiently large number (say, a million) of quantum objects, such as electrons or photons, emitted from a source, are (provisionally speaking in classical terms) allowed to pass through the slits and collide with the screen, where the traces of these collisions become recorded. We can only observe such traces as *effects* of the processes involving certain types of physical objects (quantum objects), the existence of which we infer on the basis of such traces. In other words, in each event we can only observe a mark, which we infer to be a trace left by a "collision" between a quantum object and the screen. Each such collision is similar in appearance to a collision involving a very small object, idealized as a particle in classical physics, and in their outward appearance, both cases are similar. In this sense, such individual quantum phenomena may be associated with the particle-*like* behavior of quantum objects, which need not, and in the present view does not mean, that quantum objects are particles (any more than they are waves) in the sense of classical physics.

In the first setup, both slits are open and we do not—or more significantly, in principle cannot—know which slit each quantum object passes through. In the second, we can—either in practice or, again, in principle—have such knowledge by installing devices, such as counters, which allow us to do so without appreciably disturbing the course of each individual run of the experiment (defined by an individual emission from the source). Such devices are sometimes called "which-path" or "which-way" devices. We can also close one of the slits for each

---

[3] For an excellent nontechnical but conceptually important account of quantum electrodynamics, "the strange theory of electrons and photons," see (Feynman 1985).

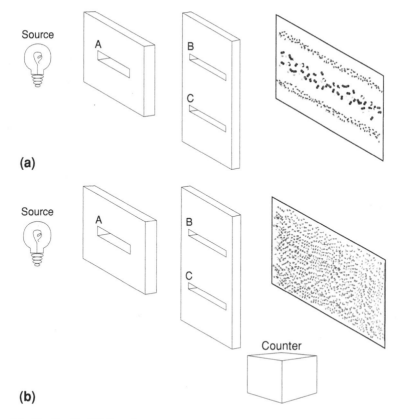

(a)

(b)

Fig. 6.1   The Double-Slit Experiment

such run, which allows each object to go through one slit only. A given quantum object could of course be blocked by the diaphragm, but these runs of the experiment are discounted. There are more or less equivalent experiments that allow us to "channel" each quantum object in a *more* controlled (it can never be fully controlled) way in each individual run of the experiment.

In the first setup, the traces of collisions between quantum objects and the screen will form a wave-*like* or, as it is usually called, "interference" pattern—a pattern similar (but, by virtue of its discrete individual constituents, not identical) to that produced by the traces of the wave processes in an appropriate medium (Fig. 6.1a). In principle (there could be practical limitations), the interference pattern will appear regardless of the distance between slits or the time interval between the emissions. This interval can be made sufficiently long for each emission to take place after the previously emitted object has reached the screen and been destroyed by its collision with the latter, which makes the appearance of the interference pattern especially remarkable and enigmatic. This interference pattern is the actual physical manifestation and, according to Bohr's and, arguably, most other interpretations, the *only* physical manifestation of quantum waves. Wave-like effects are more pronounced and more suggestive of a physical wave-propagation when we deal with very strong beams consisting of very large numbers of photons following one another in quick succession, which effects were at some point responsible for wave theories of light, culminating in Maxwell's electrodynamics. Indeed, the language of interference may not be the most suitable here. The "correlational pattern"—that of a correlated, rather than random, distribution of traces—would be a better term.

Now, in the second setup, when we install devices that allow us—either in practice or in principle—to know which slit each quantum object passes through (which can be done without appreciably disturbing the course of each run of the experiment), the wave-like interference pattern never appears (Fig. 6.1b). Accordingly, in this setup quantum objects behave in a particle-like manner both individually *and* collectively; that is, the observed random pattern of collisions is similar to that which would appear if we conducted an analogous experiment with classical objects, idealized as particles. As I said, merely setting up the apparatus in a way that would make such knowledge in principle possible, even if not actually obtained, would suffice.

The situation, sometimes known as the quantum measurement paradox, is indeed peculiar, all the more so given that the interval between emissions could be sufficiently large for the preceding quantum object to be destroyed before the next one is emitted, without affecting the appearance or lack of the interference pattern in a given setup. Other standard locutions include strange, puzzling, incomprehensible, or mysterious. The same type of features are found in other quantum experiments, which have continued to confirm this strangeness and surprise us with new, ever stranger, features of quantum behavior throughout the history of quantum physics, and they still do. Thus, how do particles "know," individually or (which may be even more disconcerting) collectively, that both slits are open and no counters are installed or, conversely, that counters are installed to check which slits particles pass through, and modify their behavior accordingly, from the wave-like to particle-like, or, another way to look at it, by changing the corresponding probability distributions? The last view of the situation is preferable, given that the observed phenomena (marks on the screen) are always particle-like; in the first setup just considered, the interference pattern only emerges out of multiple individual events (one needs about 70,000). Not everyone agrees, and this well-recognized fact does not prevent arguments to the effect that the unobserved, or even unobservable, behavior of individual quantum objects can be wave-like, specifically in the situation when the interference pattern or certain analogous phenomena appear. There is, however, no experiment that allows us to ever observe individual quantum objects as passing through both slits, and in this sense as behaving individually in the wave-like manner. I shall return to the difficulties involved in these arguments below.

The totality of the phenomena observed in the double-slit experiment is incompatible with an explanation that classical physics (that of particle-like or wave-like classical objects) can provide. Nor can classical physics properly predict the numerical data associated with these or other quantum phenomena, which is what led to quantum theory in the first place. The behavior leading to the effects observed in the double-slit experiment cannot be exhibited by the same classical entities even in different circumstances; nor can we phenomenally conceive of entities that would be simultaneously particles and waves, or continuous and discontinuous, to begin with.[4] The classical objects exhibiting the different observed behaviors leading to the two incompatible kinds of observable effects are described by two fundamentally different types of theories: by classical mechanics in the case of particle-like objects and by classical electrodynamics in the case of radiation. This is why Planck's discovery that radiation can, under certain conditions, behave in a particle-like manner was such a shock.

Indeed, the situation appears to defy any possible explanation of how quantum objects behave in space and time, and possibly (and in Bohr's ultimate view, actually) the application of the latter concepts to this behavior. Attempts to conceive of this behavior in terms of physical attributes of quantum objects themselves appear to lead to unacceptable or at least highly problematic consequences. Among such consequences are logical contradictions; a mysterious behavior of quantum objects; difficult assumptions, such as attributing volition or personification to nature in allowing particles individual or collective choices; the nonlocality of the situation, in the sense of its incompatibility with relativity; and (a form of temporal nonlocality) retroaction in time. The impossibility of understanding quantum mechanics is invoked as well, sometimes as an argument against the theory. Feynman's statement, "nobody understand quantum mechanics" (Feynman 1965, p. 128), has been endlessly recycled, usually out of context and without due attention to the situation. For one thing, as Feynman himself was well aware, the quantum situation, as manifest in the double-slit experiment, is an experimental situation and is not a product of quantum mechanics, which merely responds to it.

Admittedly, this response is seen as inadequate by some of the critics just mentioned, because it is difficult and perhaps impossible to interpret quantum mechanics as accounting for this situation by (descriptively) explaining the behavior of quantum objects leading to the phenomena observed in the double-slit and other quantum experiments. This difficulty or even impossibility especially worried Einstein, the main inspiration behind the criticisms just mentioned. Einstein's response, however, was subtler than most of those who followed him. While he initially tried to explore the apparent inconsistencies of quantum mechanics, sometimes via the type of phenomena found in the double-slit experiment, eventually he more or less accepted Bohr's view of it as logically possible and came to see quantum mechanics as an incomplete or, alternatively,

---

[4] Even in Bohmian theories, where both concepts are used in describing the behavior of quantum objects, they are not fused in a single entity: a wave accompanies and guides a particle, following de Broglie's idea, inspired by Einstein's earlier suggestion, discarded by Einstein himself.

nonlocal theory in view of the EPR-type experiments. Leaving the question of nonlocality aside for the moment (I shall address it in Chap. 8), Einstein was not wrong, if one understands by a complete theory a theory that, on the model of classical mechanics, would *describe* the independent individual behavior of quantum objects. Quantum mechanics indeed does not do so, at least in Bohr's view. As I said, the question, essentially defining the Bohr-Einstein debate, is whether nature allows us to have such a theory. Einstein thought it does or should, and Bohr thought that it just might not. Accordingly, Bohr was compelled to look for an *interpretation* of quantum phenomena and quantum mechanics, as the best available theory of these phenomena, responding, as he put it, to "the entirely new situation as regards the description of physical phenomena" (Bohr 1935, p. 700).

I shall now explain how one could consistently account for the double-slit experiment by using this interpretation, especially in its ultimate version, which suspends an assignment of any attribute, even single ones, to quantum objects. All measurable quantities considered pertain strictly to measuring instruments impacted by quantum objects. Earlier versions of Bohr's interpretation would still allow one to attribute to a quantum object itself either one or the other of conjugate quantities, such as the position or the momentum, at the time of measurement. An attribution of any quantity to quantum objects independent of measurement was always disallowed by Bohr's interpretation but, a simultaneous attribution (however made) of both conjugate quantities would be prevented by the uncertainty relations. However, Bohr was compelled to move to the ultimate version of his interpretation and the corresponding instantiation of complementarity in view of the EPR experiment, although not in his 1935 reply to EPR. As I shall indicate as I proceed, my discussion of the experiment can be adjusted, when necessary, to represent Bohr's earlier views.

Bohr was, arguably, the first to take advantage of the fact that the two setups or the two types of phenomena occurring in the double-slit experiment (or other paradigmatic quantum experiments) are always mutually exclusive. This realization allowed Bohr to contend that the features of quantum phenomena exhibited in the double-slit experiments or other key quantum experiments need not be seen as paradoxical. As he says: "It is only the circumstance that we are presented with a choice of *either* tracing the path of a particle *or* observing interference *effects*, which allows us to escape from the paradoxical necessity of concluding that the behavior of an electron or a photon should depend on the presence of a slit in the diaphragm through which it could be proven not to pass" (Bohr 1949, *PWNB* 2, p. 46). Our tracing of the path of any quantum object could, again, only amount to indirect information, and "tracing the path" (not the best expression here) only means that we can know which slits the particle has passed through, but this information is sufficient to avoid the paradoxes in question. The concept of complementarity reflects this mutual exclusivity, along with the fact, again, crucial to the architecture of the concept, that either one or the other setup can be selected.

As explained earlier, the concept of complementarity does not originate in this way, and it was originally instantiated in the Como lecture in terms of the complementarity of space–time coordination and the claim of causality. However, as explained earlier, Bohr changed his key instantiations of the concept (in part by abandoning the application of the idea of causality to quantum processes), under the impact of his initial exchanges with Einstein, in which the discussion of the double-slit experiment figured decisively. It may, accordingly, be argued, as I do here, that the double-slit experiment played a crucial role in developing Bohr's views and, specifically, in changing his use of the concept of complementarity after the Como argument, which was not grounded in the double-slit experiment. This change is clearly reflected in all of his subsequent writings, in most of which the double-slit experiment plays a central role as well. By the same token, Bohr's epistemology of quantum phenomena is essentially linked to this post-Como view of complementarity. Most especially, as discussed in the preceding chapter, the concept now becomes instantiated primarily by or in relation to the mutual exclusivity of certain experimental arrangements, such as those considered in the double-slit or most other paradigmatic quantum experiments noted above. Other key complementary relations used by Bohr, from that point on, such as that of the space–time description and the application of conservation laws are correlated with such mutually exclusive, complementary arrangements. Both types of complementary relations just mentioned are linked to the uncertainty relations (either for position and momentum or time and energy measurements). As will be seen below, the two complementary arrangements defining the double-slit experiment are correlative to the uncertainty relations, given that the one (leading to the emergence of the interference pattern) can be associated with the momentum measurement and the other (which prevents the appearance of the interference pattern) with the position measurement for each quantum object. (Each such measurement is performed when the object passes through the diaphragm with slits.)

Bohr's view of the double-slit experiment from this epistemological perspective, especially, again, in its ultimate version, may be seen as defined by several key hypotheses, which also ground his *interpretations* of quantum phenomena in general. First of all, Bohr assumes that there exist material physical systems, designated as "quantum objects," whose nature and behavior, manifested in their impact upon our measuring instruments, cannot be described by means of classical physics. He also assumes that the ultimate constituents of nature (now called "elementary particles") are quantum objects, although quantum objects in general can also be composite. That does not seem much to assume, but it is an essential starting point, also insofar as it implies that such objects exist independently of us and our measuring instruments, although there is nothing we can say about these objects apart from their interactions with measuring instruments. Accordingly, Bohr's interpretation only assumes *the existence* of quantum objects manifested in their capacity to have effects upon the world we observe, but it does not presuppose, and ultimately allow for, any further claims concerning this existence. Phenomenally or conceptually, one only deals with quantum *phenomena*, defined by the effects upon measuring instruments of the interactions between quantum objects and those instruments, in the absence of any quantum-level *specifiable* ontology, and ultimately, again, prohibiting such an ontology. On the other hand, Bohr grounds his interpretation of quantum phenomena and quantum mechanics in the specifiable classical and, hence, realist ontology of measuring instruments and the classical *epistemology* defined by this realist ontology—by what we *know* concerning the impact of quantum objects on these instruments. It need not follow that classical objects are rigorously classical at the ultimate level of their constitution, since they may be—and generally (there are exceptions) are assumed to be—ultimately quantum. Bohr's interpretation assumes that measuring instruments have quantum strata, which enables their interaction with quantum objects. Only certain strata of measuring instruments may be described classically and, as noted above, Bohr further argues that they must be so described, since it does not appear possible for us to describe our quantum experiments otherwise.[5]

Bohr also assumes that *in certain specific respects*, quantum objects individually behave physically *like* particles, while they never *individually* behave *physically* like waves or other continuously propagating (spreading) objects. In particular, keeping in mind the qualifications indicated by my emphasis and speaking provisionally of quantum objects themselves and their behavior, in the double-slit experiment a quantum object—say, an electron—never go through both slits in the wave-like manner, regardless of the setup. Each electron passes through one and only one slit, whether we do not or cannot know which slit it has passed through (which leads to the emergence of the interference pattern, once the experiment is repeated a sufficient number of times) or whether we have, or could in principle have, such knowledge (which precludes the emergence of the interference pattern).

It is sometimes argued, contrary to this view, that passing through both slits and, hence, behaving in a wave-like spreading manner is what quantum objects do in the first setup. For the reasons to be explained below, I find these arguments to be problematic. I certainly do not think that Bohr ever held this view, although it is sometimes argued that he did, mostly, it appears, because of the confusion concerning his actual argumentation rather than only a mode of expression, such as "the wave picture of the electron," sometimes found in his works, such as the Bristol lecture (Bohr 1931, p. 363). It is clear, however, that, on such occasions, specifically the one just cited, Bohr only means "symbolic waves" and uses the corresponding terms, such as "diffraction," in referring to a single electron, accordingly specifically when speaking of an electron (or photon) passing through a slit, a single slit ("hole"), as on this occasion (Bohr 1931, p. 363). In any event, Bohr never speaks of an electron and a photon passing through both slits. The point has brought home in his reply to EPR, when he explains that when a particle "impinges on the diaphragm, the diffraction by the slit of the plane wave giving the symbolic representation of its state" (Bohr 1935, p. 697). This means that the measurement de facto performed as a result gives us only probabilistic expectations concerning some future possible measurements. These expectations concern *individual* events and are defined by

---

[5] This argument relates to a thorny problem of the transition from the quantum to the classical domain—a problem addressed, with, I would argue, at most a limited success, by decoherence theories (e.g., Zurek 2003; Schlosshauer 2007).

quantum amplitudes and the rules, such as Born's rule, for quantum probabilities. The situation is correlative to the uncertainty relations, connected with this symbolic plane wave, which represents, in accordance with Born's interpretation of the wave function, the probabilistic nature of our quantum predictions *even in the case of individual quantum experiments.*

As Bohr explained in "Discussion with Einstein," before quantum mechanics, there was the following major problem. The energy and momentum of an individual particle were given, as was discovered in the case of the photon by Einstein (a revolutionary discovery in its own right), as respectively $E = h\nu$ and $P = h\sigma$, where "h is Planck's constant, and ν and σ are a number of vibrations per unit time and the number of waves per unit length" (Bohr 1949, *PWNB* 2, p. 33). The same type of formulas were eventually proposed by de Broglie for the electron and quickly found valid. Bohr used them, as *symbolic* formulas, throughout his writings, such as the Bristol lecture and his reply to EPR. As Bohr says: "Notwithstanding its fertility, the idea of the photon implied a quite unforeseen dilemma, since any simple corpuscular picture of radiation would obviously be irreconcilable with interference effects, which present so essential an aspect of radiative phenomena, and which can be described only in terms of a wave picture. The acuteness of the dilemma is stressed by the fact that the interference effects offer our only means of defining the concepts of frequency and wave-length entering into the very expressions for the energy and momentum of the photon" (Bohr 1949, *PWNB*2, p. 34). Or so they did before these "wave" terms were understood in terms of the wave function and its probabilistic interpretation, which made the use of these terms "symbolic."

Now, even though I state more unequivocally that photons *never* individually behave physically like waves, and make only a qualified appeal to photons' particle-*like* individual behavior, both claims require further qualifications. For one thing, they amount to at least a partial assessment concerning how quantum objects behave or (as will be seen, this difference is important here) at least *do not behave* apart from measurements, which is hazardous in quantum theory and is prohibited in Bohr's interpretation. A rejection of the idea of quantum waves as *physical* waves or, at least, qualifications concerning the use of such a concept are common, although the idea is not dead, even beyond the Bohmian theories, in which quantum waves are given a physical meaning, along with and alongside particles. However, qualifications that must be made concerning the particle-*like* behavior of quantum objects, while less common, are no less important. First of all, we do not appear to be able—nobody has ever accomplished this thus far—to observe, through any instrument, the independent behavior, say, motion (particle-like or wave-like), of quantum objects. We can only observe certain trace-like effects of this behavior manifested in these instruments, and we infer the existence of quantum objects from these effects. Quantum objects themselves are usually destroyed in the process of this "irreversible amplification" of their quantum interaction with measuring instruments to the classical level (Bohr 1949, *PWNB* 2, p. 51; Bohr 1958, *PWNB* 3, p. 3). These effects, such as a trace on a silver bromide screen or a click in a detector, define, in Bohr's terms, *individual* quantum phenomena. These phenomena are always particle-*like* insofar as such individual traces, form contained, point-*like*, individual entities, always discrete relative to each other.[6]

Accordingly, the statement "a photon passed through a slit" only means that a measuring device registered an event that is *analogous* to a certain classical physical event, say, that of the hitting of a screen by a small classical object that passed through an opening in some diaphragm on its way. The statement "a photon never passes through both slits" means that no event corresponding to such a statement can be observed or registered. We can never register an individual event simultaneously linked to both slits, say, by placing a detector near each slit. Only one of these detectors registers each individual event: the two detectors never click simultaneously. This fact already poses considerable difficulties for the assumption that a photon can pass through both slits. These difficulties are amplified by other considerations, in particular, a potential retroaction in time, in view of Wheeler's delayed choice experiment, which allows us to decide on the type of setup we want to use *after* each object or even all of them have already passed through the slits (Wheeler 1983, pp. 190–192).[7] On the other hand, one could speak of a single photon as "passing through a slit" in the sense that the corresponding event could be registered by a "which-path" measuring device, but only in this sense. The very difficulty of conceiving of the independent behavior of quantum objects is in part due to the apparent change in their behavior depending on the measuring arrangements, such as their "propensity" to fit into collective patterns in some, but only some, arrangements.

In sum, either type of characterization—particle-*like* (which can be both individual and collective) and wave-*like* (which is only collective)—only relates to the behavior of quantum objects as concerns the effects of this behavior on measuring instruments, or phenomena in Bohr's sense, since we observe nothing else. Neither concept—that of "wave" or that of "particle"—applies as a physical concept to quantum objects and their behavior themselves. The individual phenomenal *effects* in question in the double-slit experiment and other quantum experiments may be

---

[6] These traces are not really "points": they appear as "dots" only at a low resolution, while actually comprising millions of atoms (Ulfbeck and Bohr 2001; Bohr, Mottelson and Ulfbeck 2004). These authors target the use of the idea of particles, including by Niels Bohr (one of the authors is Aage Bohr, Niels Bohr's son), in my view, mistakenly. In particular, they disregard Bohr's argument that the concept of particle, just as that of wave, is an abstraction and (this is crucial) an abstraction of a different kind than in classical physics.

[7] Wheeler appears to assume that each photon passes through both slits in the corresponding setup, or more accurately makes an equivalent assumption in his experiment, which is a version of the beam-splitter experiment. That Wheeler subscribes to the idea is intriguing because he is among the stronger advocates of Bohr's views, which appear to be in conflict with this idea. As explained above, Bohr sometimes applies the term "wave" to individual particles in the symbolic sense. I do not believe, however, that statements to the effect that individual electrons or photons actually, physically behave in a wave-like manner (or pass through more than one slit) ever occur in Bohr. On the other hand, statements at least suggesting the opposite are found throughout his writings (e.g., Bohr 1949, *PWNB* 2, pp. 46–47; Bohr 1935, p. 697).

seen as *particle-like* insofar as they are *similar* to the kind of traces classical particle-like objects colliding with the screen would leave as well. One cannot, however, automatically infer from this similarity that quantum objects are particle-like objects of the type we deal with in classical physics. Quantum objects certainly do not behave in the way particles do in classical physics, any more than in the way classical waves do. In particular, because of the uncertainty relations, a quantum object cannot be simultaneously assigned, as, at least ideally, an object can in classical physics, both an exact position and an exact momentum, and hence a trajectory—the difficulty that became apparent early in the history of quantum theory.[8] In Bohr's ultimate interpretation of the situation, we cannot ever assign even a single such property to a quantum object, which view goes beyond the uncertainty relations but is obviously consistent with them. Nor can we apply to quantum objects classical physical concepts associated with these properties, such as those of motion, or even use words such as "happens" or "occurs." As both Bohr and Heisenberg argue, such words can only apply at the level of observation, and not to what happens before an observation or between observations and hence not to quantum objects themselves (Bohr 1949, *PWNB* 2, pp. 50–51; Heisenberg 1962, pp. 51–58).

Bohr, accordingly, sees the quantum–mechanical situation as indicating "the ambiguity in ascribing customary physical attributes to atomic [quantum] objects" themselves or to their independent behavior, as against phenomena in his sense, something that is actually observed or registered (Bohr 1949, *PWNB* 2, p. 51). This perspective allows him to give a consistent view of the double-slit experiment and related experiments in terms of the complementary nature of the phenomena considered in these experiments. He writes: "To my mind, there is no other alternative than to admit that, in this field of experience, we are dealing with individual phenomena and that our possibilities of handling the measuring instruments allow us only to make a choice between the different complementary phenomena we want to study" (Bohr 1949, *PWNB* 2, p. 51). Thus, the interference pattern only reflects the correlationally ordered distribution of the traces left by photons in the first setup, as opposed to a different distribution of such traces found in the second setup. In other words, we are dealing not with properties of quantum

---

[8] In quantum theory, elementary particles are idealized as zero-dimensional, point-like objects, even though they are assigned masses. Such objects can be given a rigorous mathematical meaning, but not a rigorous physical meaning. It is true that this type of point-like (mathematical) idealization of physical objects is also used in classical physics. There, however, this idealization allows one to approximate the actual behavior of the objects considered and, on the basis of this descriptive approximation, to make excellent predictions concerning this actual behavior. The physical objects themselves thus considered may be, and usually are, assumed to have extension, the property that has defined physical objects or, more generally, material bodies (*res extensa*) at least since Descartes. In the case of quantum objects, such an assumption is difficult to sustain. For example, in the case of electrons it leads to a well-known contradiction with classical electrodynamics, since, if assumed to have extension, an electron would be torn apart by its negative charge. This circumstance led to the idealization of the electron as a dimensionless point-like object even before quantum mechanics.

objects but with two different and mutually exclusive types of (individual) observable or registered effects upon measuring instruments of the interaction between quantum objects and those instruments under particular, rigorously specified physical conditions. By the same token, it is *not our knowledge of the behavior of quantum objects* but *our knowledge concerning classical physical events registered in measuring instruments* that defines the absence or the appearance of the interference pattern in the double-slit experiment.

6.3 The Uncertainty Relations, Complementarity, and Probability

The double-slit experiment is essentially connected to both the uncertainty relations and the irreducibly probabilistic nature of our quantum predictions, which are in effect correlative to each other. This three-faceted situation became crucial to Bohr's thinking and was at the core of his debate with Einstein from 1927 on. The uncertainty relations, $\Delta q \Delta p \cong h$, numerically represent the insuperable limits on the simultaneous determination, by means of either measurement or prediction, of both the position and the momentum (or certain other pairs of variables, such as time and energy) associated with a quantum object.[9] It is an experimental fact that both quantities can never be measured simultaneously beyond the limits of accuracy defined by h, regardless of the precision of our instruments, although, as I shall explain presently, the physical meaning of this statement is far from straightforward, and is a matter of interpretation. I would like to note, first, that, although one can derive, as Bohr did in the Como lecture, the uncertainty relations more immediately and in a more elementary way by using de Broglie's formulas for matter waves, Heisenberg originally derived them from quantum mechanics, via Dirac's transformation theory (Heisenberg 1927). That he could do so has a special significance. On the one hand, as was shown by Heisenberg's γ-ray-microscope thought experiment in his paper and by Bohr's derivation in the Como lecture, the uncertainty relations could be seen as experimentally given, a law of nature, and they were so seen by both Heisenberg and Bohr. On the other hand, Heisenberg's derivation showed that quantum mechanics "contains" the uncertainty relations. In other words, it showed that quantum mechanics adequately responds to the experimental data in question. As will be seen in Chaps. 8 and 9, this situation became important for Bohr's argumentation in his debate with Einstein, who saw the uncertainty relations as an artifact of quantum mechanics

---

[9] As I noted earlier, we measure the momentum in a given direction, and the uncertainty relations apply to this momentum and the corresponding coordinate. In the uncertainty relations for the position and the momentum associated with a quantum object in 3D space, each quantity will have three components defined by the chosen coordinate system.

and sought to circumvent them in his arguments that aimed to show the incompleteness or else nonlocality of quantum mechanics.

The physical meaning of Heisenberg's formula itself is, again, a subtle matter, and I shall adopt the following view, which is courtesy of Asher Peres, but is consistent with Bohr's view:

> The only correct interpretation of [an uncertainty relation $\Delta x \Delta p \cong h$, where x is a coordinate and p the momentum in the same direction] is the following: If the *same* preparation procedure is repeated many times, and is followed either by a measurement of x, or by a measurement of p, the various results obtained for x and for p have standard deviations, Δx and Δp, whose product cannot be less than [h]. There never is any question here that a measurement of x "disturbs" the value of p and vice versa, as sometimes claimed. These measurements are indeed incompatible, but they are performed on *different* [*quantum objects*] (all of which are identically prepared [in terms of the physical state of the measuring instruments]) and therefore these measurements cannot disturb each other in any way. The uncertainty relation [$\Delta x \Delta p \cong h$] only reflects the intrinsic randomness of the outcomes of quantum tests. (Peres 1993, p. 93)

Not everyone would subscribe to Peres's claim that this is "the only correct interpretation" of the uncertainty relations or to this interpretation itself. For my purposes, it suffices that this interpretation is consistent with the experimental evidence and with Bohr's views. As Peres also observes, "an uncertainty relation [such as $\Delta q \Delta p \cong h$] is not a statement about the accuracy of our measuring instruments. On the contrary, its derivation assumes the existence of *perfect* instruments" (Peres 1993, p. 93). Bohr corroborates this observation in a striking sentence that brings in the role of measuring instruments, which define *phenomena*, as different from quantum *objects*, and defines the uncertainty relations themselves via the complementary characters of space–time concepts and conservation laws. He says "we are of course not concerned with a restriction as to the accuracy of measurement, but with a limitation of the well-defined application of space–time concepts and dynamical conservation laws, entailed by the necessary distinction between measuring instruments and atomic objects" (Bohr 1958, *PWNB* 3, p. 5, also Bohr 1937, *PNNB* 4, p. 86; Bohr 1954, *PWNB* 2, p. 73).

As indicated earlier, the emergence of the interference pattern in the double-slit experiments can be properly correlated with the possibility of the (ideally) precise *momentum* measurement (at the diaphragm with slits) for each quantum object involved in the corresponding setup, and the lack of the interference pattern with the possibility of the (ideally) precise *position* measurement (at the slit) associated with each quantum object in the alternative setup. Since the two setups are always mutually exclusive or complementary both quantities cannot be measured in the same experiment. One must be careful in applying Bohr's concept of complementarity to collective, rather than individual, phenomena, but there is no difficulty in this context, since each measurement involved is associated with a single quantum object and a single observed phenomenon, associated with the corresponding measuring arrangement. This connection between the double-slit experiment and the uncertainty relations is used by Bohr throughout his exchanges with Einstein in order to counter-argue Einstein's criticism of quantum mechanics from 1927 on (e.g. Bohr 1935, pp. 607–700; Bohr 1949, *PWNB* 2, pp. 41–47, 52–61).

In his account of his 1927 exchange with Einstein in "Discussion with Einstein," Bohr explains his use of the double-slit experiment to address "the question of whether the quantum–mechanical description exhausted the possibilities of accounting for observable phenomena, or, as Einstein maintained, the analysis could be carried further and, especially, of whether a fuller description of the phenomena could be obtained by bringing into consideration the detailed balance of energy and momentum in *individual processes*" (Bohr 1949; *PWNB* 2, pp. 42-43; emphasis added). The trend of Einstein's thought, to be continued into all of his criticism of quantum mechanics, is clearly apparent here. Quantum mechanics is a correct but incomplete theory, and the uncertainty relations reflect this incompleteness. Bohr counterargued that the uncertainty relations are uncircumventable, a law of nature, at least as things stood then and as they still stand now (Bohr 1949, *PWNB* 2, pp. 42–47). Accordingly, quantum mechanics does exhaust "the possibilities of accounting for observable [quantum] phenomena," and hence is complete. That is, it is as complete as nature allows our theory to be, although under these circumstances one can be careful as to the meaning of the phrase "the quantum–mechanical description," since quantum mechanics does not provide the description of individual quantum processes. It is not a "mechanics" of such processes in the sense classical mechanics is. In this respect, Einstein was right. As I have stressed throughout this study, the question is whether nature allows for such a mechanics or, as Bohr argues, whether it might not and, at least, again, as things stand now, does not. In other words, contrary to Einstein's view or hope, it may not be possibly to carry further "the analysis of [quantum] phenomena." As Bohr says, again, in "Discussion with Einstein," by now commenting on Einstein's later (1936) argument, "in quantum mechanics we are not dealing with an arbitrary renunciation of a more detailed analysis of atomic phenomena, but with a recognition that such an analysis is *in principle* excluded" (Bohr 1949, *PWNB* 2, p. 62–63). Einstein, as Bohr notes on the same occasion, rejected this argumentation on principle, at least as a basis on which future fundamental theories should be built. By that time (1936), Einstein, who apparently accepted Bohr's earlier counterarguments, came up with a new alternative, manifest, or so he believed, in the EPR and related experiments. He now argued that quantum mechanics may be a complete theory, but then it is nonlocal, insofar as it implies, as he called, "a spooky action at a distance" (which would be in conflict with relativity). I shall discuss this argument of Einstein and Bohr's responses to them in Chap. 8. My main point here that in all of the Bohr-Einstein exchanges, including those around the EPR and related experiments, the double-slit experiment plays a key role in Bohr's argumentation.

The situation that obtains in the double-slit experiment is also equivalent to the probabilistic and (statistically) correlational nature of quantum predictions. These correlations are manifested in the interference pattern setup of the experiment— although that pattern, and hence the corresponding correlational order, are, again, formed by an accumulation of *random* individual events. The "history" of any single event as such can never be certain, and no single run of the experiment is ever guaranteed to be repeatable. A single event registered by a counter cannot be used to establish unconditionally that an object passed through a slit, any more than can any given trace on the screen. These circumstances reflect one of the greatest mysteries of quantum phenomena, perhaps their greatest ultimate mystery: How events that are irreducibly random can, under certain circumstances, give rise to order, even if only a correlational order. Thus, the double-slit experiment and the uncertainty relations both, and correlatively, reflect the probabilistic and correlational order found in quantum phenomena, and hence the enigmatic relationships between chance and probability in quantum mechanics, which predicts the probabilities in question in exact correspondence with experiments.

That does not, again, mean that alternative accounts of quantum phenomena are not possible, although descriptive accounts appear difficult to achieve. Be that as it may, Bohr's interpretation of quantum phenomena and of quantum mechanics responds both to the conceptual difficulties of the situation and to the fact that quantum mechanics properly predicts all of its numerical aspects. It does so by suspending or even forbidding, in principle, the possibility of knowing how such or any other effects of quantum objects upon our world (classically observed by us) are possible; or, correlatively, the possibility of knowing what happens to quantum objects between the experiments. More radically, it precludes any knowledge or even conceptualization of quantum objects and processes. In this respect, quantum mechanics may be even more incomplete than Einstein argues. For it is not only a matter of obtaining at most partial information concerning quantum objects and their behavior—for example, knowing only the position and not the momentum of a given quantum object at a given point, in accordance with the uncertainty relations (if one applies them to quantum objects, rather than to measuring instruments). There is no knowledge available and ultimately no conceptualization (beginning with conceiving of them as quantum or objects in any specific sense) possible concerning quantum objects and their behavior. There is only knowledge (fundamentally probabilistic in nature when it comes to our predictions) concerning the effects of quantum objects upon our measuring instruments or other macro-objects in the (classical) world we observe, conceptualize, and know.

Bohr's interpretation does not resolve the great enigmas of quantum physics: What are quantum objects? What is the nature of their independent behavior? How are the observable features of quantum phenomena possible? How can the irreducibly random character of individual events coexist with the correlational patterns at the collective level found in certain experimental setups? The significance of Bohr's interpretation is that it tells us that this enigma may be irresolvable: perhaps no explanation or spatio-temporal conceptualization of the behavior of quantum objects will ever be possible. This impossibility, however, opens the way to doing so much physics that would be impossible otherwise.

Chapter 7
1933. "On the Question of Measurability of Electromagnetic Field Quantities": Complementarity and Quantum Field Theory

7.1 Quantum Field Theory in Bohr and Beyond

This chapter addresses the subject that has, as yet, been little explored in literature on Bohr—the significance of quantum field theory, beginning with quantum electrodynamics (the first form of quantum field theory), in Bohr's work on complementarity. This significance is considerable, in particular, in the following three respects. *First*, Bohr saw quantum electrodynamics and other forms of quantum field theory as confirming his key ideas concerning the epistemology of quantum phenomena and quantum mechanics, and possibly giving these ideas more radical dimensions through the new mathematics and physics these theories introduce. *Second*, by extending his thinking concerning quantum mechanics to these theories, Bohr made important contributions to our understanding of measurement in quantum field theory in his influential collaborations with Léon Rosenfeld in 1933 and 1950. *Third*, quantum electrodynamics and quantum field theory had a shaping reciprocal impact on Bohr's work on quantum mechanics and complementarity.

I shall proceed as follows. This section serves as a general introduction to quantum electrodynamics and quantum field theory. Section 7.2 offers a discussion of some of the key aspects of Bohr's collaboration with Rosenfeld, "On the Question of Measurability of Electromagnetic Field Quantities," and its implication for the epistemology of quantum electrodynamics and quantum field theory. Section 7.3 will outline the epistemological situation at stake in quantum field theory in more general conceptual terms.

Most of Bohr's major works on quantum mechanics and complementarity contain important references to quantum electrodynamics and quantum field theory. This should hardly come as a surprise. The birth of quantum electrodynamics was nearly simultaneous with that of quantum mechanics. It is almost certain that Schrödinger wrote a relativistic wave equation for the electron before discovering his nonrelativistic equation. That equation, now known as the Klein-Gordon equation, was abandoned by Schrödinger as unworkable, although it was eventually proven to be not

A. Plotnitsky, *Niels Bohr and Complementarity*, SpringerBriefs in Physics,
DOI: 10.1007/978-1-4614-4517-3_7, © Arkady Plotnitsky 2013

without merits in quantum electrodynamics and was also used elsewhere in quantum field theory. The relativistic (free) electron is described, in a particle form, by Dirac's equation, discovered in 1928. A preliminary discussion of quantum electrodynamics (due to Jordan) appeared already in the three-man paper on matrix mechanics (Born et al. 1926). Most relevant to Bohr's thinking were the early work on quantum electrodynamics by Jordan and Dirac, and then Dirac's relativistic equation for the electron. Jordan and Dirac were also cofounders of the transformation theory, which connects Heisenberg's and Schrödinger's versions of quantum mechanics. As discussed earlier, Dirac's paper on the subject (Dirac 1927) was written in Copenhagen, and had a significant influence on Bohr's thinking in the Como lecture, although, as also explained earlier, the corresponding epistemological aspects of the Como argument were abandoned by Bohr. Dirac's discovery of his equation of the electron in 1928 was an example of his masterful use of the transformation theory, one of his favorite tools. Dirac's first paper on quantum electrodynamics was written, while at Bohr's institute in Copenhagen in 1926, at the time of Bohr's work leading to complementarity and Heisenberg's work leading to the uncertainty relations, and their famously heated exchanges on quantum mechanics and its interpretation, mentioned in Chap. 3. Several other physicists who were at Bohr's Institute and some of whom were Bohr's assistants, such as Hendrik Kramers, Pauli, Heisenberg, Oscar Klein, and Léon Rosenfeld, made major contributions to quantum electrodynamics and quantum field theory. An important exchange of letters between Bohr and Dirac occurred at the time of Dirac's work on his positron theory, following the discovery of his equation, which predicted the existence of the positron. Dirac's letter announces ideas that have had a lasting impact on the subsequent development of quantum theory (Letter to Bohr August 10, 1933).[1] Bohr continued to be actively involved in the discussions concerning quantum field theory, especially the emerging theory of nuclear forces, in the 1930s.[2]

Bohr's most important direct intervention into quantum field theory was his collaboration with Rosenfeld on the question of measurement in quantum field theory, first in "On the Question of Measurability of Electromagnetic Field Quantities" (Bohr and Rosenfeld 1933) and then in "Field and Charge Measurement in Quantum Electrodynamics" (Bohr and Rosenfeld 1950). The second article was largely based on the argument of the first paper, but was written in the wake of important new developments in quantum electrodynamics in the 1940s, developments to be discussed in Sect. 7.2. "On the Question of Measurability of Electromagnetic Field Quantities" was written in response to Lev Landau and Rudolf Peierls's argument (Landau and Peierls 1933). The exchange, which had preceded EPR's argument and Bohr's reply by only 2 years, involved major

---

[1] A copy of the letter is in the Niels Bohr Library, American Institute of Physics, New York. See A. Pais's discussion of this letter in (Pais 1986, pp. 382–383). These ideas, I argue here, had a significant impact on Bohr's view of quantum field theory.

[2] Pais's discussion of quantum field theory and nuclear physics during the 1930s gives a clear sense of Bohr's engagements with it (Pais 1986, pp. 296–438), as does his discussion of this part of Bohr's work in his biography of Bohr (Pais 1991, pp. 346–374).

physical and epistemological issues concerning the similarities and differences between quantum mechanics and quantum electrodynamics or quantum field theory, in particular the question of the uncertainty relations, which Landau and Peierls claimed to be no longer applicable in quantum field theory. Bohr and Rosenfeld argued the contrary. Bohr's work on the nature of measurement in the case of (quantum) electromagnetic field variables was clearly important to his thinking concerning quantum measurement in quantum theory in general. The same thinking shaped Bohr's argument, based on the examination of quantum measurement, in his reply to EPR, although EPR's paper and Bohr's reply dealt with quantum mechanics.[3]

Quantum electrodynamics, the quantum theory of electromagnetism, and quantum field theory deal with quantum processes of high energy, when the effects of Einstein's (special) relativity theory must be expressly taken into account. This makes both theories relativistic, in contrast to quantum mechanics, where such effects can be neglected because the speed of the objects considered by it is slow (vis-à-vis the speed of light) and the energy involved is lower. As will be discussed in next chapter, quantum mechanics remains *consistent* with relativity as concerns all predictions it makes. Other quantum field theories are the theory of strong nuclear forces, eventually developed into the so-called quantum chromodynamics (dealing with quarks and gluons) and the theory of weak nuclear forces, which was unified with quantum electrodynamics a few decades ago. These theories now form the so-called standard model. Quantum field theory serves as an umbrella form for all these theories, and it will be used here in this way as well, unless qualified otherwise, and such qualifications are sometimes necessary.[4]

Quantum electrodynamics was introduced by Dirac in 1926–1927 as a theory in which both electrons and photons were treated in terms of particles, and his famous relativistic equation for the electron, discovered in 1928, followed the same approach. By contrast, Heisenberg and Pauli developed a quantum field theory, by analogy with classical electrodynamics, a classical field theory of electromagnetic radiation, which was introduced, in the nineteenth century, by James C. Maxwell and which was based on a new concept, that of field, developed by Michael Faraday and Maxwell. The corresponding concept in quantum field theory is that of "quantum field," which I shall explain below. A little later, in 1930, Enrico Fermi introduced a version of quantum field theory in which the photons were treated in

---

[3] Rosenfeld was Bohr's assistant at the time, and he specifically assisted Bohr in writing his reply to EPR. Bohr's and Rosenfeld's respective interpretations of quantum mechanics are, however, different, and one should treat with caution Rosenfeld's comments on Bohr's views.

[4] The philosophy of quantum field theory is a complex and far from sufficiently developed subject, in comparison with quantum mechanics, as the paucity of literature addressing the subject, as against that on quantum mechanics, suggests. One can think of barely a handful of books devoted to the subject in contrast to the unending stream of books on the philosophy of quantum mechanics. One might mention (Cao 2004) and Paul Teller's *Quantum Field Theory: An Interpretive Introduction* (Teller 1995), which also contains useful further references. For historical accounts, see (Schweber 1994) and, more technical, (Weinberg 2005). For a more accessible account see (Feynman 1985).

terms of (quantum) fields and the electrons in terms of particles. Eventually, it became clear any form of "quantum field theory" (using it as an umbrella term) could be presented mathematically equivalently either in terms of "particles" or in terms of "fields," in (qualified) parallel with "particle" and "wave" in quantum mechanics. As in quantum mechanics, at least, in Bohr's interpretation, physical terms, such as waves, particles, and fields (already a more complex entity as far as its physical meaning is concerned in classical electromagnetism) could only be used provisionally or by a symbolic analogy.

Let us recall that, in developing quantum mechanics, first Heisenberg and then Born and Jordan formally adopted the equations that would describe the motion of particles in classical mechanics, most generally, in their Hamiltonian form, but gave these equations a new physical content by using mathematically different (matrix) variables, to which the equations themselves applied. As a result, quantum mechanics would only predict, in general probabilistically, the outcomes of the corresponding experiments, without describing the properties of quantum objects and their behaviors in the way classical mechanics describes the behavior of classical physical objects. In developing his (particle-like) quantum electrodynamics, including in deriving his famous equation, Dirac adopted the same approach, which he had already used earlier in creating his own version of quantum mechanics, taking as inspiration Heisenberg's original paper, but introducing even more general variables, "q-numbers," as he called them. In the case of quantum electrodynamics, he developed a (Hamiltonian) formalism applied to this type of variable, analogous to but quite a bit more complex than in quantum mechanics, based on the mathematics of the so-called spinors, which form what is known as Clifford algebras. Bohr had reflected on this point already in the Como lecture (in the published version, since the Como conference itself preceded Dirac's discovery). Bohr said: "Dirac has been able successfully to attack the problem of the magnetic electron through a new ingenious extension of the *symbolical* method and so to satisfy the relativity requirement without abandoning the agreement with spectral evidence. In this attack not only the imaginary complex quantities appearing in the earlier procedures are involved, but the fundamental equations themselves contains quantities of a still higher degree of complexity that are represented by matrices" (Bohr 1927, *PWNB* 1, p. 90).

In their version of quantum field theory Heisenberg and Pauli used the field picture. In this respect it was the first rigorous quantum *field* theory, since, as a particle theory, Dirac's electrodynamics could be seen as a form of *mechanics*. Heisenberg and Pauli's approach was similar to that of Schrödinger in developing his wave mechanics, but, unlike Schrödinger, Heisenberg and Pauli suspended the physical picture of wave propagation, or of classical field, or all classical-like physical pictures, from the outset. In the classical electromagnetic theory of Faraday and Maxwell, the concept of field associates with each point of a propagating "field" a vector or a set of vectors, representing actual physical forces, active at this point, which gives rise to a kind of geometrical picture of the field. Nothing like that was possible in Heisenberg and Pauli's theory, in this respect extending, just as Dirac did, the approach Heisenberg used in creating quantum mechanics, in contrast to Schrödinger's visualizable (*anschaulich*) wave program.[5] Heisenberg and Pauli used equations symbolically analogous to the (wave) equations of classical electromagnetism, Maxwell's equations, to develop the mathematics of their theory, but again, using new quantum types of variables. These equations, too, only predicted the probabilities of the outcomes of the experiments in question. Later on Heisenberg, Fermi, and Hideki Yukawa introduced the quantum field theory of nuclear (weak and strong) forces.

Quantum field theory has remained a theory of both "particles" and "fields," allowing one to use either "picture," or either form of the "unpicturable," or to variously combine both, depending on one's need or preference. In this respect, the situation is, again, analogous to the one that obtains in quantum mechanics, where one could use either Schrödinger's "wave" equation or a more algebraic formalism of matrix mechanics, or that of transformation theory that combines both.[6] The complementarity of particles and fields is sometimes invoked as well. It is, however, different from the wave-particle complementarity of quantum mechanics, and requires an even greater caution in using any physical analogies with classical physics. One might say that the complementarity of particles and fields is more mathematically defined, although, in the present view, neither complementarity, nor again, the concepts of particles, waves, or fields, to begin with, are ultimately applicable to quantum objects themselves and their behavior. In his comments on quantum field theory, Bohr never invokes the complementarity of particles and fields. But then, as I explained, he does not invoke wave-particle complementarity either. The primary reason in both cases is, again, that neither complementarity corresponds to the mutually exclusive individual situations of measurements, which, I argue here, define the primary forms of complementarity for Bohr, following the Como lecture.

These considerations explain why Bohr argued that quantum field theory may be seen as retaining the key epistemological features of quantum mechanics. Both quantum mechanics and quantum field theory describe the same quantum objects in the first place, if in different circumstances, defined by the levels of energy at which the corresponding processes take place. It would, accordingly, be difficult to expect that these objects would behave differently, although the interpretation of this behavior and the corresponding observable phenomena, or of quantum theories themselves, may be and has been a matter of debate, equally in both regimes. It is also not inconceivable that the current form of quantum theory, from quantum mechanics to quantum field theory and beyond, will be replaced by an epistemologically more classical theory, as Einstein hoped, although, as I said, a move

---

[5] Schrödinger's equation itself could be seen in "particle-like" terms, a point noticed, with some surprise, by Schrödinger himself, who was also one of the first to discover the mathematical equivalence of his wave mechanics and matrix mechanics.

[6] By now, in the wake of von Neumann's work, a more unified formalism of Hilbert-space or an even more abstract type (e.g., that of C*-algebras) is generally used, which allows one even greater flexibility and effectiveness, and the same is true in quantum field theory.

toward an epistemologically more radical theory is not inconceivable either, and Bohr thought it to be more likely. It is true that some of these objects, such as those (quarks and gluons) found inside nuclei or W and Z bosons mediating electroweak interactions, only appear at very high energies and, hence, can only be handled by quantum field theory, but this does not undermine my main point here. For one thing, photons, too, require quantum electrodynamics to be properly considered as quantum objects.

Bohr described Dirac's theory as "a most striking illustration of the power and fertility of the general quantum-mechanical way of description" and as reflecting "new fundamental features of atomicity" (Bohr 1949, *PWNB* 2, p. 63). The term "atomicity" is used here in Bohr's special sense, roughly equivalent to Bohr's concept of phenomena and is to be discussed in Chap. 9. This comment occurs in "Discussion with Einstein," published in 1949, by which time Bohr's interpretation of quantum mechanics and his instantiations of complementarity are based on this concept, developed by Bohr in the wake of EPR's and related arguments by Einstein. Bohr's general point, however, is that, while quantum field theory moves beyond quantum mechanics, it retains "the power and fertility of the *general* quantum-mechanical way of description" and its fundamental features and concepts, on which it builds. The theory, thus, could be seen, or at least consistently interpreted, as conforming to Bohr's epistemology of quantum mechanics, and thus showing the possibilities of new theories under these conditions. Bohr commented on this situation in 1958, clearly with Dirac's theory in mind, as a paradigm for future developments of quantum theory, which he (rightly) expected to be necessary. He says: "Such argumentation [i.e., that defined by his ultimate epistemology] does of course not imply that, in atomic physics, we have no more to learn as regards experimental evidence and the mathematical tools appropriate to its comprehension. In fact, it seems likely that the introduction of still further abstractions into the formalism will be required to account for the novel features revealed by the explorations of atomic processes of very high energy" (Bohr 1958, *PWNB* 3, p. 6). The subsequent developments of quantum field theory, which rely on ever more abstract and complex mathematics, have proven Bohr to be right.

7.2 Quantum Field Theory and Measurements

The new complexities of quantum field theory appear to be deeply connected to the question of measurement, which is so essential to Bohr's interpretation of quantum mechanics, and which takes on new dimensions in quantum field theory, beginning with quantum electrodynamics. Bohr addresses the subject at the early stages of the development of the theory in the 1930s. One of the key features of low energy quantum phenomena, on which our predictions, such as those made by means of quantum mechanics, depend, is that we can disregard the quantum aspects of the constitution of measuring instruments in defining quantum phenomena, although these aspects are responsible for the emergence of these phenomena. As Bohr explains: "although, of course, the existence of the quantum of action is ultimately responsible for the properties of the materials of which the measuring instruments are built and on which the functioning of the recording devices depends, this circumstance is not relevant for the problem of the adequacy and completeness of the quantum-mechanical description in its aspects here discussed [i.e., as concerns what can be actually observed and predicted]" (Bohr 1949, *PWNB* 2, p. 51). However, as Bohr noted in 1937, the situation acquires new complexities in quantum field theory where the quantum constitution of measuring instruments might need to be taken into account. He says:

> On closer consideration, the present formulation of quantum mechanics in spite of its great fruitfulness would yet seem to be no more than a first step in the necessary generalization of the classical mode of description, justified only by the possibility of disregarding in its domain of application the atomic structure of the measuring instruments themselves in the interpretation of the results of experiment. For a correlation of still deeper laws of nature involving not only the mutual interaction of the so-called elementary constituents of matter but also the stability of their existence, this last assumption can no longer be maintained, as we must be prepared for a more comprehensive generalization of the complementary mode of description which will demand a still more radical renunciation of the usual claims of the so-called visualization. (Bohr 1937, *PWNB* 4, p. 88)

Bohr's view here expressed can be related to Pauli's argument that in the case of low-energy quantum phenomena (treated by quantum mechanics) the observer is still too "detached" or even "too completely detached" from quantum objects (Pauli 1994, p. 132). Disregarding "the atomic structure of the measuring instruments" is primarily responsible for this detachment. Given the ultimate form of Bohr's epistemology of quantum mechanics, (which was just about in place at this point), it is difficult to think of "a still more radical renunciation" than the one defined by this epistemology. It may, however, be argued that quantum field theory gives further impetus to the argument for this renunciation, and perhaps additionally pushed Bohr toward it. At the same time, quantum-field-theoretical phenomena introduced a more radical form of multiplicity of phenomena than those found in quantum-mechanical phenomena into quantum field theory. I shall discuss this aspect of quantum field theory in Sect. 7.3.

For the moment, Bohr's observation just cited must have been brought about, at least in part, by Bohr's work with Rosenfeld on measurement in quantum field theory in response to Landau and Peierls's argument (Bohr and Rosenfeld 1933). The latter argument, contested by Bohr and Rosenfeld, concerned a possible inapplicability of the uncertainty relations in quantum field theory. I shall, however, by and large bypass this subject. My main concern here is the conceptual architecture that responds to the question of measurement in quantum field theory in view of the possible role of "the atomic structure of the measuring instruments" in quantum-field-theoretical measurement vis-à-vis quantum-mechanical measurement, where this structure plays no role. As Bohr and Rosenfeld's paper demonstrates, the probabilistic relationships between the formalism of quantum field theory and measurements analogous to those that obtained in quantum mechanics could still be maintained, even though at a further cost, relative to that paid in quantum mechanics. According to Bohr and Rosenfeld:

> [I]t is also of essential importance that the customary description of an electric field in terms of the field components at each space–time point, which characterizes classical field theory and according to which the field should be measurable by means of point charges in the sense of the electron theory, is an idealization which has only a restricted applicability in quantum theory. This circumstance finds its proper expression in the quantum-elec-tromagnetic formalism, in which the field quantities are no longer represented by true point functions but by functions of space–time regions, which formally correspond to the average values of the idealized field components over the regions in question. The for-malism only allows the derivation of unambiguous predictions about the measurability of such region-functions, and our task will thus consist in investigating whether the com-plementary limitations on the measurability of field quantities, defined in this way, are in accordance with the physical possibilities of measurement. Insofar as we can disregard all restrictions arising from the atomistic structure of the measuring instruments, it is actually possible to demonstrate a complete accord in this respect.
>
> Besides a thorough investigation of the construction and handling of the test bodies, this demonstration requires; however, consideration of certain new features of the comple-mentary mode of description, which come to light in the discussion of the measurability question, but which were not included in the customary formulation of the indeterminacy principle in connection with non-relativistic quantum mechanics. Not only is it an essential complication of the problem of field measurements that, when comparing field averages over different space–time regions, we cannot in an unambiguous way speak about a temporal sequence of the measurement processes; but even the interpretation of individual measurement results requires a still greater caution in the case of field mea-surements than in the usual quantum-mechanical measurement problem. (Bohr and Ro-senfeld 1933, pp. 480–481)

One might note, first, that, as will be seen in the next chapter, the trend of thought and the mode of expression of Bohr's reply to EPR is clearly felt here. In both cases, the argumentations offered are based on the careful examination of what can and cannot be unambiguously ascertained in quantum mechanics and quantum field theory, vis-à-vis classical physics (respectively, classical mechanics or classical electrodynamics); in both cases, it is also argued that neither EPR nor Landau and Peierls sufficiently pursue this type of examination. The main point at the moment is that the measurements of quantum fields involve additional com-plexities of idealization related to the question of "the atomic structure of the field sources and the measuring instrument." Bohr and Rosenfeld's conclusion in their article brings this point into a sharper focus. As they write:

> We thus have arrive at the conclusion already stated at the beginning, that with respect to the measurability question the quantum theory of fields represents a consistent idealization to the extent that we can disregard all limitations due to the atomic structure of the field sources and the measuring instruments. .... [T]his result should properly be regarded as an immediate consequence of the fact that both the [quantum-electromagnetic] formalism and the viewpoints on which the possibilities of testing this formalism are to be assessed have as their common foundation the correspondence argument. Nevertheless, it would seem that the somewhat complicated character of the considerations used to demonstrate the agreement between formalism and measurability are hardly avoidable. For in the first

place the physical requirements to be imposed on the measuring arrangement are condi-
tioned by the integral form in which the assertions of the quantum-electromagnetic formalism are expressed, whereby the peculiar simplicity of the classical field theory as a
purely differential theory is lost. Furthermore, as we have seen, the interpretation of the
measuring results and their utilization by means of the formalism require consideration of
certain features of the complementary mode of description which do not appear in the
measurement problems of non-relativistic quantum mechanics. (Bohr and Rosenfeld 1933,
pp. 520–521)

This assessment has proven to be prescient when assessed against the subsequent development of quantum electrodynamics and quantum field theory, and
specifically attempts to resolve certain problems of the theory. These problems,
however, had begun to emerge by the time of Bohr and Rosenfeld's paper. They
have primarily to do with the appearance of infinities or divergences in the theory,
once one attempted to use the theory to make calculations that would provide
closer approximations matching certain experimentally observed data.

These difficulties were eventually resolved through the so-called renormalization procedure, which became and has been ever since a crucial part of the
machinery of quantum electrodynamics and quantum field theory. In the case of
quantum electrodynamics, renormalization was performed in the 1940s by Tomonaga, Schwinger, and Feynman, which brought them a joint Nobel Prize in
1965, with some contribution by others, especially Dyson, and earlier Hans Bethe
and Hendrik Kramers. (We have encountered Kramers earlier in this study as
Bohr's assistant in Copenhagen.) The Yang-Mills theory, which grounds the
standard model, was eventually shown to be renormalizable as well, by Martinus
Veltman and Gerardus t'Hooft in the 1970s (eventually bringing them their Nobel
Prize). This allowed a proper development of the standard model of all forces of
nature, except for gravity, which has not, as yet, been given its quantum form.

The procedure is extremely difficult mathematically, and while there are ways
to see it in more benign ways, its mathematical legitimacy is still not entirely
established. Roughly speaking, the procedure might be seen as manipulating
infinite integrals that are divergent and, hence, mathematically illegitimate. At a
certain stage of calculation, however, these integrals are replaced by finite integrals through artificial cutoffs that have no proper mathematical justification within
the formalism and are performed by putting in, by hand, experimentally obtained
numbers that make these integrals finite, which removes the infinities from the
final results of calculations. These calculations are experimentally confirmed to a
very high degree. Indeed, quantum electrodynamics is the best experimentally
confirmed theory in our possession. I cannot address the subject of renormalization
in detail, and can only give a brief summary of what is at stake courtesy of Dyson.[7]
According to Dyson:

----

[7] See (Teller, pp 149–168) and, for a historical account, (Schweber 1994, pp. 595–605). More
recent developments, such as renormalization groups, effective quantum field theory, and so forth
cannot be discussed here, in part because of their nearly prohibitive technical aspects. They also
do not appear to me to change the epistemological argument offered in this chapter.

[Quantum fluctuations of the electromagnetic field in the atom, say, the hydrogen atom] would give the electron an additional energy, called the self-energy. It was well known that [Dirac's] quantum electrodynamics (QED) gave an infinite value for the self-energy and was therefore useless. Physics has reached an impasse. On the one hand, the Lamb experiment gave clear evidence that the effects of electromagnetic quantum fluctuations were real and finite. On the other hand, the existing theory of QED gave infinite and absurd results....

[In the1940s, however, Kramers] remarked that the observed energy of an electron, according to QED is the sum of two unobservable quantities: a bare energy, which the electron is supposed to have when it is uncoupled from electromagnetic fields, and the self-energy, which results from the electromagnetic coupling. The bare energy appears in the equations of the theory but is physically meaningless, since the electromagnetic coupling cannot really be switched off. Only the observed energy is physically meaningful. The point of renormalization was to get rid of bare energies and replace them with observed energies. (Dyson 2005, p. 48)

This was what was essentially accomplished, thus restoring legitimacy to Dirac's theory, rather than abandoning it and replacing it with a different theory, as some, Dirac, the founder of the theory, among them, thought necessary. It is clear from these comments that the role of observation and measurement has essential bearings on the situation, although the questions thus posed remain formidable and are far from fully resolved even now. One might ask, however, why, say, quantum electrodynamics (to which I shall limit my discussion at the moment, although the situation is the same in quantum field theory) contains its infinities in the first place, in contrast to quantum mechanics? Bohr's argumentation, arising from the analysis given in his paper with Rosenfeld, suggests a possible answer, which is quite subtle and is roughly as follows. The infinities and, hence, the necessity of renormalization appear to arise because the mathematical formalism of quantum electrodynamics is essentially linked or even based on the quantum-mechanical idealization of measurement of the type invoked by Bohr and Rosenfeld, an idealization that disregards "the atomic structure of the field source and the measuring instruments." A better idealization and the corresponding mathematical formalism, which could avoid infinities, should be based on an idealization that would respond, however indirectly, to "the atomic structure of the field source and the measuring instruments." Such an idealization is currently unavailable, but it may be possible, as an alternative *predictive* formalism, to which Bohr's epistemology of quantum phenomena limits us, since this epistemology in principle precludes a descriptive analysis of this atomic structure. According to Dyson:

We interpret the contrast between the divergent Hamiltonian formalism [which imposes the necessity of renormalization] and the finite S-matrix as a contrast between two pictures of the world, seen by two observers having a different choice of measuring equipment at their disposal. The first picture is of a collection of quantized fields with localizable interactions, and is seen by a fictitious observer whose apparatus has no atomic structure and whose measurements are limited in accuracy only by the existence of the fundamental constants c and h. This ["ideal"] observer is able to make with complete freedom on a sub-microscopic scale the kind of observations which Bohr and Rosenfeld employ ... in their classical discussion of the measurability of field-quantities. The second picture is of collection of observable quantities (in the terminology of Heisenberg) and is the picture seen by a real observer, whose apparatus consists of atoms and elementary particles and whose measurements are limited in accuracy not only by c and h, but also by other constants such as α [the fine-structure constant] and m [the mass of the electron]. (Dyson 1949, p. 1755, cited in Schweber 1994, pp. 547–548)

S. Schweber's commentary on Dyson's paper further clarifies the situation:

A "real observer" can measure energy levels, and perform experiments involving the scattering of various elementary particles—the observables of S-matrix theory—but cannot measure field strengths in small regions of spacetime. The "ideal" observer, making use of the kind of "ideal" apparatus described by Bohr and Rosenfeld, can make measurements of this last kind, and the commutation relations of the fields can be interpreted in terms of such measurements. The Hamiltonian density will presumably always remain unobservable to the real observer whereas the ideal observer, "using nonatomic apparatus whose location in space and time is known with infinite precision[,]" is presumed to be able to measure the interaction's Hamiltonian density. "In conformity with the Heisenberg uncertainty principle, it can perhaps be considered a physical consequence of the infinitely precise knowledge of location allowed to the ideal observer, that the value obtained by him when he measures Hamiltonian density is infinity" (Dyson 1949, p. 1755). If this analysis is correct, Dyson speculated, *the divergences of QED are directly attributable "to the fact that the Hamiltonian formalism is based upon an idealized conceptualization of measurability."* (Dyson 1949, p. 1755) (Schweber 1994, p. 548; emphasis added)

This is possible. Hamiltonian formalism, initially brought into quantum mechanics, via the (mathematical) correspondence principle, from classical physics, was transferred by Dirac into quantum electrodynamics, via the correspondence of its predictions to those of quantum mechanics at the quantum-mechanical limit, since at this limit Dirac's equation converts itself into Schrödinger's equation. As explained above, both earlier theories did not need to take into account the atomic structure of measuring instruments or sources of fields, classical physics naturally and quantum mechanics, it appears, without real justification, but, luckily, still allowing us to make correct predictions concerning the outcomes of quantum experiments. With quantum electrodynamics we run out of luck as concerns the formalism, which may need to be based otherwise to avoid divergences and the necessity of renormalization. We are still lucky, however, because renormalization works thus far, at least insofar as our quantum theories can disregard gravity. It is, accordingly, not surprising that Bohr thought that mathematical abstractions beyond those of current quantum field theory might have been necessary, at least if one is to avoid renormalization, which Bohr probably had in mind when he made his 1958 comment to that effect, cited above. This is especially likely given that his second collaboration with Rosenfeld in 1950 expressly referred to the corresponding developments in quantum electrodynamics (Bohr and Rosenfeld 1950, p. 523). Dyson makes a similar point concerning a possible alternative to the standard Hamiltonian formalism of quantum electrodynamics (Dyson 1949, p. 1755). It may be added that when Bohr made his 1958 comment, the quantum field theory of nuclear forces had not yet been renormalized, and it was by no means clear that it could be, and in the way that it was by Veltman and t'Hooft in the 1970s came somewhat as a surprise. So Bohr might have thought that an alternative mathematical future quantum theory would be required in any event, whether a finite one or one requiring renormalization.

It is difficult to know how this situation will play itself out in the future, for example, when quantum gravity will be developed, assuming it will be. Our future theories might be finite (some versions of string and brane theory appear to hold such a promise), thus proving that the necessity of renormalization is merely the result of the limited reach of our quantum theories at present. Beginning with quantum electrodynamics, quantum field theory confronts a much more complex manifold of quantum phenomena than does quantum mechanics. This complexity may in turn arise because the quantum constitution of measuring instruments comes into play more manifestly in these phenomena than in those considered by quantum mechanics. While a finite theory may be preferable, renormalization may not be a very big price to pay for the theory's extraordinary capacity to predict these phenomena, without, again, describing quantum objects themselves and their behavior. This epistemology makes the recourse to probability in such predictions unavoidable in principle, as it is in quantum mechanics. This recourse is also unavoidable in practice, given that in both cases identically prepared experiments in general lead to different outcomes. As discussed earlier, however, that does not automatically exclude a descriptive theory of the corresponding phenomena, in the way it would in an interpretation consistent with Bohr's epistemology of quantum mechanics. For Bohr, however, quantum field theory provides, to return to his phrase describing quantum mechanics, "mathematical instruments," to handle the unthinkable nature of quantum objects and processes that manifest their existence, although not their character, in the new types of effects observed in measuring instruments at higher energy levels.

7.3 How Many Particles?

As discussed in Chap. 3, the development of quantum mechanics by Heisenberg was defined by three key elements, which were immediately grasped by Bohr and which shaped his own thinking concerning quantum mechanics and complementarity throughout. The same three elements also defined the discovery of quantum electrodynamics by Dirac, especially in his derivation of his relativistic equation for the electron. Dirac's thinking on his equation was influenced by Heisenberg's thinking in his first paper of matrix mechanics. It is well-known that Dirac had read the paper very carefully earlier, and that it had inspired his own work on quantum mechanics and his earlier papers on quantum electrodynamics, from which, especially from his transformation theory, his work on his equation grew. It is true that Dirac's derivation of his equation also followed and depended on Schrödinger's equation, and on transformation theory. Nevertheless, Heisenberg's thinking, discussed in Chap. 3, continued to exert a profound influence on Dirac's work on his equation. Especially significant are the following three key points:

(1) *The Mathematical Correspondence Principle* Stemming from Bohr's correspondence principle, the mathematical correspondence principle requires recovering the equations and variables of the old theory, classical mechanics in the case of quantum mechanics and quantum mechanics in the case of quantum electrodynamics, in the limit region where the old theory applies. In Heisenberg's and Dirac's work alike (keeping the difference of the limit theory in mind), the principle served as an important guidance for developing a new theory.

(2) *The Introduction of New Types of Variables* Arguably most centrally, both discoveries, that of Heisenberg and that of Dirac, were characterized by the introduction of new types of *mathematical variables*, or one might say, in parallel with Newton's discovery of classical mechanics, a new *calculus* of quantum theory:

(QM) In the case of quantum mechanics, these were matrix variables with *complex* coefficients, essentially operators in Hilbert spaces over *complex* numbers, vs. classical physical variables, which are differential functions of *real* variables. Heisenberg formally retained the equations of classical physics themselves.

(QED) In the case of quantum electrodynamics, these were Dirac's spinors and multi-component wave functions, which, jointly, form more complex operator variables, reflecting a more complex structure of the corresponding Hilbert space of the theory, again, over complex numbers. In contrast to Heisenberg, Dirac also introduced a new type of equation, which is different from Schrödinger's equation, but is, by the mathematical correspondence principle, consistent with the latter, as a far non-relativistic limit of Dirac's theory, via Pauli's spin-matrix theory, which is the immediate non-relativistic limit of Dirac's theory.

(3) *The Probabilistically Predictive Character of the Theory* This change in *mathematical* variables was accompanied by a fundamental change in *physics* vis-à-vis classical physics or relativity: the variables and equations of quantum mechanics and quantum electrodynamics no longer describe, *even by way of idealization*, the properties and behavior of quantum objects themselves. Instead, these theories only predict, in general probabilistically, the outcomes of quantum events and statistical correlations between some of these events.

Heisenberg's revolutionary thinking established a new way of doing theoretical physics, and, as a consequence, this thinking ultimately redefined experimental physics as well. The practice of *experimental physics* no longer consists, as in classical experiments, in tracking the independent behavior of systems considered, but in *unavoidably creating* configurations of experimental technology that reflect the fact that what happens is *unavoidably defined* by what kinds of experiments we perform, how we affect quantum objects, rather than only by their independent behavior. My emphasis on "unavoidably" reflects the fact that, while the behavior of classical physical objects is sometimes affected by the corresponding configurations of experimental technology used in classical experiments, in general we can observe classical physical objects, say, planets moving around the sun, without appreciably affecting their behavior. This does not appear to be possible in quantum experiments.

In view of these circumstances, the practice of *theoretical physics* no longer consists, as in classical physics or relativity, in offering an idealized mathematical description of quantum objects and their behavior. Instead it consists in developing mathematical machinery that is only able to predict, in general (in accordance with what obtains in experiments) probabilistically, the outcomes of events and of correlations between some of these events.

The situation takes an even more radical form in quantum field theory and experimental physics in the corresponding (high) energy regimes. This became quickly apparent, following the invention of quantum electrodynamics and especially Dirac's equation, albeit not quite immediately, since the radical character of the situation, beginning with the existence of antimatter, took a while to realize. The situation itself extends to all forms of quantum field theory. While fully retaining the epistemology just outlined, quantum field theory, beginning with quantum electrodynamics, is characterized by, correlatively:

(1) more complex configurations of phenomena observed and hence of the measuring apparatuses involved (including modern-day particle accelerators), and thus more complex configurations of the effects of the interaction between quantum objects and measuring instruments;
(2) a more complex nature of the mathematical formalism of theory, reflected in the necessity of renormalization;
(3) a more complex character of quantum-field-theoretical predictions and, hence, of the relationships between the mathematical formalism and the measuring instruments involved in high-energy physics.

To illustrate these complexities, suppose that one arranges for an emission of an electron, at a given high energy, from a source and then performs a measurement at a certain distance from that source. Placing a photographic plate at this point would do. The probability of the outcome would be properly predicted by quantum electrodynamics. But what will be the outcome? The answer is not what our classical or even our quantum-mechanical intuition would expect, and this unexpected answer was a revolutionary discovery of quantum electrodynamics. Let us consider, first, what happens if we deal with a classical and then a quantum object in the same type of arrangement.

Consider, as a model of the classical situation, a small ball that hits a metal plate, which can be considered as either a position or a momentum measurement, or indeed a simultaneous measurement of both, and time t. In classical mechanics we can deal directly with the objects involved, rather than with their effects upon measuring instruments. The place of the collision could, at least in an idealized representation of the situation, be predicted exactly by classical mechanics, and we can repeat the experiment with the same outcome on an identical or even the same object. Most importantly, regardless of where we place the plate, we always find the same object, at least in a well-defined experimental situation, which is shielded from significant outside interferences, such as, for example, those that can deflect or even destroy the ball earlier.

By contrast, if we deal with an electron as a quantum object in the quantum-mechanical (low-energy) regime we cannot predict the place of collision exactly and, correlatively, exactly repeat the experiment on the same electron. Also, correlatively, we cannot simultaneously predict, or measure, the position and the momentum of an electron, which makes the situation subject to the uncertainty relations, or the uncertainty relations a reflection of the situation. Indeed, there is a nonzero probability that we will not observe such a collision at all, or that if we do, that a different electron (coming from somewhere else) is involved. It is also not possible to distinguish two observed traces as belonging to two difference objects of the same type. Unlike in the classical case, in dealing with quantum objects (which, thus, cannot be identified with phenomena, as they can be in classical physics), there is no way to improve the conditions or the precision of the experiment to avoid this situation. Quantum mechanics, however, gives us correct probabilities for such events. This is accomplished by defining the corresponding Hilbert space, with the position and other operators as "observables," writing down Schrödinger's equation, and using Bohr's rule or similar rules, to obtain the probabilities of possible outcomes.

Once the process occurs at a high energy and is governed by quantum electrodynamics, the situation is still different, even radically different. One might find, in the corresponding region, not only an electron, as in classical physics, or an electron or nothing, as in the quantum-mechanical regime, but also other particles: a positron, a photon, an electron–positron pair. Just as does quantum mechanics, quantum electrodynamics, beginning with Dirac's equation, rigorously predicts which among such events can or cannot occur, and with what probability, and, in the present view it can only predict such probabilities or, sometimes, statistical correlations between quantum events. In order to do so, however, the corresponding Hilbert-space machinery becomes much more complex, essentially (speaking for the mathematically inclined audience) making the wave function ψ a four-component Hilbert-space vector, as opposed to a one-component Hilbert-space vector, as in quantum mechanics. I cannot enter technical mathematical details here, and they are not necessary to make the main point: this more complex mathematical architecture naturally allows for a more complex structure of predictions (which are, again, probabilistic) corresponding to the situation just explained, usually considered in terms of the so-called virtual particle formation and Feynman's diagrams.

Once we move to still higher energies or different domains governed by quantum field theory the panoply of possible outcomes becomes much greater. The corresponding Hilbert spaces would be given a yet more complex structure, properly correlated with different elementary particles, interactions, and so forth. In the case of Dirac's equation we only have electron, positron, and photon.

It follows that in quantum field theory an investigation of a particular type of quantum object irreducibly involves not only other particles of the same type but also other *types* of particles. This qualification is important, because, as I said, the identity of particles within each type is strictly maintained in quantum field theory, just as it is in quantum mechanics, and in either theory one cannot distinguish different particles of the same type, such as electrons. One can never be certain that one encounters the same electron in a given experiment in the quantum-mechanical situation, although the probability that it would in fact be a different electron is low in the quantum-mechanical regime in comparison to that in the regime of quantum electrodynamics. In quantum field theory, it is as if instead of identifiable moving objects and motions of the type studied in classical physics we encounter a continuous emergence and disappearance, creation and annihilation, of particles from point to point, theoretically governed by the concept of virtual particle formation. The corresponding operators acting in Hilbert spaces, which are used to predict the probability of such events, are called the creation and annihilation operators. The particular possible events are usually represented now in terms of Feynman diagrams. This takes us beyond quantum mechanics. For, while the latter questions the applicability of such classical concepts as objects (particles or waves) and motion, and possibly all concepts, at the quantum level, it still preserves the identity of quantum objects.

The introduction of a new mathematical formalism, still of a Hilbert-space type, but involving more complex Hilbert spaces, and operator algebras, was a momentous event in the history of quantum physics, comparable to that of Heisenberg's introduction of his matrix variables. It is difficult to overestimate the significance of this mathematical architecture, which, most famously, amounted to the discovery of antimatter, and more generally, led to a very radical view of matter, which the existence of antimatter reflects.

Bohr's assessment of Dirac's theory, mentioned earlier, was also prompted by these considerations, in addition to the role of the quantum structure of measuring instruments, as discussed earlier. Bohr expressly refers to these considerations in "Discussion with Einstein":

> Dirac's ingenious quantum theory of the electron offered a most striking illustration of the power and fertility of the general quantum-mechanical way of description. In the phenomena of creation and annihilation of electron pairs we have in fact to do with new fundamental features of atomicity, which are intimately connected with the non-classical aspects of quantum statistics expressed in the exclusion principle, and which have demanded a still more far-reaching renunciation of explanation in terms of a pictorial representation. (Bohr 1949, *PWNB* 2, p. 63)

It is, again, not clear how much further one can renounce "explanation in terms of a pictorial representation," once such a representation or any representation or even conception is already renounced altogether, as it is in Bohr's ultimate interpretation of quantum mechanics, in place by the time of this remark. One might, however, understand why Bohr thought that Dirac's theory and, following it, all quantum field theory make a return to classical-like epistemology all the more unlikely. "Atomicity" should be understood here in Bohr's sense, a concept that, as will be discussed in Chap. 9, governs his ultimate view of quantum phenomena and quantum mechanics. In higher-energy regimes, described by quantum electrodynamics and quantum field theory, "atomicity" acquires a more radical form. This is because, as explained above, one could no longer count on the stable identity of elementary constituents of nature, even in considering the "individual" behavior of individual objects, such as electrons, photons, etc., because such processes always involve multiple types of particles.

Accordingly, while, in Bohr's view, we are still dealing with the effects of the interactions between quantum objects and measuring instruments, we now deal with a greater multiplicity of such effects and more complex rules for probabilities for the events (effects) involved. Each actual, registered event in the quantum-field-theoretical picture could be described in terms of such effects, and, hence, phenomena in Bohr's sense, while virtual events relate to possible phenomena. In reflecting on the situation in the early 1970s, Heisenberg, too, was compelled to see Dirac's discovery of anti-particles, a discovery that first revealed this situation, as "perhaps the biggest of all the big changes in physics of our century ... because it changed our whole picture of matter" (Heisenberg 1989, pp. 31–32).

The theory made remarkable progress and has acquired a much richer content and structure since its introduction or since Heisenberg's remark, as is manifest most famously in the electroweak unification and the quark model of nuclear forces, developments that commenced around the time of these remarks. Many predictions of the theory, from quarks to electroweak bosons and the concept of confinement and asymptotic freedom, to name just a few, were spectacular, and, since its introduction, the field has garnered arguably the greatest number of Nobel Prizes in physics. It was also quantum field theory that led to string and then brane theories, the current stratosphere of theoretical physics. However, the essential epistemological points in question have remained in place, just as is the case with the epistemology of quantum mechanics, although, as stressed throughout this book, the debates concerning them and Bohr's epistemology have never subsided or lost any of their intensity.

In this respect, Bohr's and Heisenberg's assessments just cited would only require relatively minor adjustments. A more recent statement by Pais, who was both a major practitioner of the theory and a major historian of the subject, confirms this. The statement is, it is true, no longer that recent either. It was made in 1986, but not much has changed in this respect since, as is clear, for example, from Frank Wilczek's 2005 review of the present state of the theory (Wilczek 2005), or from any number of more recent assessments of the situation, although technical achievements of the theory (or relevant experimental physics) continue to remain momentous. Wilczek received his 2004 Nobel Prize (shared with David Gross and David Politzer) for his contribution, although the work itself (on the so-called asymptotic freedom of quarks), was done before Pais's comments and is discussed in Pais's book. According to Pais:

> Is there a theoretical framework for describing how particles are made and how they vanish? There is: quantum field theory. It is a language, a technique, for calculating the probabilities of creation, annihilation, scattering of all sorts of particles: photons, electrons, positrons, protons, mesons, others, by methods which to date invariably have the

character of successful approximations. No rigorous expression for the probability of any of the above-mentioned processes has ever been obtained. The same is true for the corrections, demanded by quantum field theory, for the positions of energy levels of bound-state systems. There is still a Schrödinger equation for the hydrogen atom, but it is no longer exactly soluble in quantum field theory. In fact, ... the hydrogen atom can no longer be considered to consist of just one proton and one electron. Rather it contains infinitely many particles.

In quantum field theory the postulates of special relativity and of quantum mechanics are taken over unaltered, and brought to a synthesis which perhaps is not yet perfect but which indubitably constitutes a definitive step forward. It is also a theory which so far has not yielded to attempts at unifying the axioms of general relativity with those of quantum mechanics. Is quantum field theory the ultimate framework for understanding the structure of matter and the description of elementary processes? Perhaps, perhaps not. (Pais 1986, p. 325)

In Bohr's view, one would, again, not be able to speak rigorously of "the *description* of elementary processes," since this inability would itself appear as a postulate of quantum mechanics in Bohr's interpretation, taken over unaltered by quantum field theory. Pais does not make strong claims to that effect in the book either. It is worth noting that Pais was Bohr's assistant in the 1940s, and he wrote a more or less definitive biography of Bohr (Pais 1991), as well as that of Einstein (Pais 1982). As these books make clear, Pais was sympathetic to Bohr's philosophy of physics (vis-à-vis that of Einstein), although he does not appear to follow Bohr's epistemology to its ultimate limits.

Quantum field theory gives further supports to Bohr's argument for the epistemology of quantum phenomena and quantum theory, and they contain new epistemological complexities that appear likely to take them further in the direction defined by this epistemology. It is, again, possible that quantum field theory will, as Einstein hoped, eventually be developed on epistemologically more classical lines, or replaced by other theory that is epistemologically more classical, even within the present scope of quantum field theory, and hence, apart from the fact that the theory does not incorporate gravity. Given the current state of quantum theory and the experimental evidence, it is difficult to predict how things will turn out on that score, or even to expect that we will arrive at a definitive answer any time soon. One thing is just about certain, however: whatever epistemology, whether closer to that advocated by Bohr or to that desired by Einstein, will win the day next time around, some remarkable physics is likely to emerge.

Chapter 8
1935. "Can Quantum-Mechanical Description of Physical Reality Be Considered Complete?": The EPR Experiment and Complementarity

8.1 From BKS to EPR

This chapter offers an analysis of Bohr's exchanges with Einstein concerning the completeness and locality of quantum mechanics, most especially Einstein, Podolsky, and Rosen's article and Bohr's reply, both published under the same title—"Can Quantum-Mechanical Description of Physical Reality Be Considered Complete?"—in 1935. EPR's article introduced a thought experiment, the EPR experiment, and offered a particular argument concerning it, EPR's argument (to be distinguished from the EPR experiment), which led EPR to conclude that quantum mechanics is incomplete, or else nonlocal, in the sense of entailing instantaneous physical connections between events, connections forbidden by relativity. These conclusions were questioned by Bohr in his reply, which offers a different analysis of the EPR experiment and derives different conclusions concerning its meaning and implications. While important for Bohr and Einstein (and a few others at the time), the EPR experiment and the Bohr–Einstein exchange concerning it had been somewhat marginal to the debate concerning quantum theory, until Bell's and related theorems discovered in the 1960s and then the experimental findings associated with these theorems. These developments have brought the EPR experiment to center stage of this debate, indeed largely defined now by the key questions posed by the experiment, especially those concerning the locality of quantum phenomena and quantum mechanics.

At the same time, Bohr's reply remains, arguably, the least understood and the most contested work considered in these discussions. Accordingly, I felt compelled to offer a more extended analysis of the exchange and, especially, of Bohr's reply in this chapter. After a general introduction given in this section, Sects. 8.2 and 8.3 focus on those parts of, respectively, EPR's and Bohr's arguments that concern the completeness of quantum mechanics. Section 8.4 discusses the question of locality in both arguments. Section 8.5 is a "postscript" on earlier exchanges between Bohr and Einstein, discussed by Bohr in "Discussion with Einstein," his most sustained account of these exchanges, considered there,

A. Plotnitsky, *Niels Bohr and Complementarity*, SpringerBriefs in Physics, DOI: 10.1007/978-1-4614-4517-3_8, © Arkady Plotnitsky 2013

however, through the prism of his exchange with EPR and even his ultimate interpretation of quantum phenomena and quantum mechanics.

In the Spring of 1925, Bohr wrote in a letter to Heisenberg: "I am forcing myself these days with all my strengths to familiarize myself with the mysticism of nature and am attempting to prepare myself for all eventualities, indeed even for the assumption of a coupling of quantum processes in separated atoms. However, the costs of this assumption are so great that they cannot be estimated within the ordinary space–time description" (Letter to Heisenberg, 18 April 1925, Bohr 1972–1999, vol 5, pp. 79–80; cited in Mehra and Rechenberg 2001, vol 6, p. 163). This assessment is linked to the subject at the core of the dilemma posed by the EPR experiment, which is not surprising in retrospect, given that the experiment reflects some of the deeper "mysteries" of quantum physics. What makes Bohr's statement remarkable is that it was made in 1925. This is not only 10 years before EPR's article introducing the EPR experiment, or even before Bohr's earlier exchanges with Einstein in 1927 but also before (albeit by only a few months) Heisenberg's discovery of quantum mechanics, based on abandoning "the ordinary space–time description." The immediate context of the statement is the collapse of the Bohr, Kramers, and Slater (BKS) theory, in view of the recently obtained (by Geiger and Bothe) experimental evidence in favor of the exact validity of the energy conservation law. As noted earlier, the BKS theory was based on the statistical rather than exact conservation of energy, which did not endear it to most physicists at the time. The theory also represented the last-ditch attempt on Bohr's part to retain, partially, the space–time description of quantum processes in the manner of the old quantum theory. Heisenberg, the recipient of the letter, did not address the question of "a coupling of quantum processes in distant atoms" in his work on quantum mechanics. However, he was ready to give up, "the ordinary space–time description," which he was able to do while preserving the exact energy conservation. At the same time, his new theory retained what proved to be a lasting contribution of the BKS theory: the probabilistic nature of our predictions concerning the outcome of primitive individual quantum processes and events. These features of quantum phenomena and quantum mechanics eventually enabled Bohr to avoid physically coupling separate quantum processes in the case of the EPR experiment.

Thus, already in 1925 Bohr was contemplating both the possibility that *quantum phenomena* could exhibit the coupling features found in the situations of the EPR-type and the assumption "that [quantum processes] cannot be estimated within the ordinary space–time description," the assumption that came to define his ultimate epistemology of quantum phenomena. It is not clear whether Bohr contemplated the alternative, which defined his post-EPR views, between nonlocality (under the realist assumption of the possibility of such a description) and the lack of realism at the time, but it is not inconceivable that he did. In any event, unlike Einstein, Bohr never appears to have thought that the EPR experiment would imply the nonlocality of quantum phenomena or quantum mechanics. Just as did Einstein, Bohr saw nonlocality as unacceptable, although Einstein and others thought (mistakenly) that Bohr allowed for it in order to argue that quantum mechanics is complete. For Bohr, the cost, which, unlike Einstein, he was willing to pay, of the completeness of quantum mechanics was the impossibility of "the ordinary space–time description" and, correlatively, the irreducibly probabilistic nature of our predictions concerning even individual quantum processes and events.

The argument of Bohr's reply proved difficult to formulate. As he said in 1949: "Rereading these passages, I am deeply aware of the inefficiency of expression which must have made it very difficult to appreciate the trend of the argumentation aiming to bring out the essential ambiguity involved in a reference to physical attributes of objects when dealing with phenomena where no sharp separation can be made between the objects themselves and their interaction with the measuring instruments" (Bohr 1949, *PWNB* 2, p. 61). In part because of these difficulties, it is a matter of interpretation whether Bohr's reply offered a fully developed counterargument to EPR's argument, and that it did is still in dispute. I shall argue here that, at the very least, Bohr's reply supplied just about all the ingredients necessary for and came very close to formulating such a counterargument.[1]

It might be useful, before I proceed to my discussion of the Bohr-EPR exchange, to briefly reflect on the current stage of the debate concerning quantum phenomena and quantum mechanics, given that this stage is, as I said, largely defined by the key questions posed by the EPR experiment, most especially the role of quantum correlations. These correlations can be ascertained experimentally, independently of any theory. Quantum mechanics, however, properly predicts the corresponding numerical data, and this capacity is reflected in the concept of "entangled states" in the formalism of the theory. EPR's article has only de facto introduced quantum correlations or quantum entanglement, which reflects these correlations in the formalism of quantum mechanics. Neither term was invoked and neither concept considered in EPR's article or in Bohr's reply. The concept of "entanglement" [*Verschränkung*] was introduced by Schrödinger in his analysis of the EPR experiment (Schrödinger 1935, p. 161). The subject of correlations took center stage with Bell's theorem, based on David Bohm's version of the EPR experiment for spin measurements and discovered in the mid-1960s. Most of these developments deal with Bohm's version of the experiment, in part because, while the original thought experiment proposed by EPR is an idealized experiment that cannot be actually performed in a laboratory, Bohm's version can be. Among the most prominent of these developments, apart from Bell's theorem, are the Kochen–Specker theorem or the more recent theorems of D. M. Greenberger, M. Horne, and A. Zeilinger and L. Hardy, and, from the experimental side, A. Aspect's experiments and related experimental work, such as that of A. Zeilinger and his group (Greenberger et al. 1989, 1990; Hardy 1993; Aspect et al. 1982). The advent of quantum information theory during the last two decades gave this problematic further prominence (e.g., Jaeger 2007; Mermin 2007).

---

[1] It is not possible, within the scope of this study, to offer a close reading of Bohr's reply itself, which would be necessary in order to make the strongest possible case for my claim. I do, however, offer such a reading in (Plotnitsky 2009, pp. 279–312).

The EPR correlations are essentially quantum (classical phenomena do not exhibit them) and, like all quantum-mechanical regularities, they are statistical. As stressed throughout this study, it is a defining feature of quantum phenomena that identically prepared quantum experiments (as concerns the state of the measuring instruments involved) in general lead to different outcomes. This circumstance makes the recourse to probability irreducible in estimating the outcomes of individual experiments or events. At the same time, however, series of certain quantum events are correlated, as against analogous series of spatially separated random events found in classical physics, as is to be expected following Einstein's relativity and the principle of local causality it introduced. It is, accordingly, not surprising that the shift of attention to the subject of entanglement and correlations has brought more sharply into focus the question of locality, as against that of the completeness of quantum mechanics.[2] The currently prevalent view appears to be that quantum mechanics is complete within its proper scope. The completeness of quantum mechanics was, as the identical title of the two papers states, the primary focus of the Bohr–Einstein exchange concerning the EPR experiment. However, the question of locality was at stake in the exchange as well. As do later arguments by Einstein (e.g., 1936, 1948, 1948b, and 1949a), EPR's argument contends that quantum mechanics is *either incomplete or nonlocal*. Einstein maintained this view until the end of his life. As he said in 1947: "I cannot seriously believe in [quantum mechanics] because the theory cannot be reconciled with the idea that physics should represent a reality in time and space, free from spooky actions at a distance" (Letter to Born, 3 March, 1947, Born 2005, p. 155). Since Einstein saw nonlocality as impermissible, he never accepted that quantum mechanics could be regarded as a complete theory. The phrase "spooky actions at a distance" has been famous ever since.

As will be discussed in Chap.9, Einstein admitted that quantum mechanics could be considered local if seen as a theory of ensembles, loosely on the model of classical statistical physics. This view, however, still leaves it incomplete either by virtue of not offering a complete description of the individual quantum systems (because in this case the EPR-type arguments still apply in Einstein's view) or by offering no such description at all. Bohr agrees that quantum mechanics does not describe individual quantum processes in space and time as independent of their interaction with measuring instruments in the way classical mechanics does. Bohr argued, however, that the probabilistic character of quantum-mechanical predictions concerning single individual phenomena and events is not incompatible with the completeness of quantum mechanics because this character may be unavoidable given the nature of these phenomena. Nor, he further argued, is the situation

---

[2] As throughout this study, I use the term "local" in the sense of compatibility with (special) relativity. One encounters a number of conceptions of locality or nonlocality in the discussions of the EPR-type experiments. Other terms, such as nonseparability, are used and are given various meanings in turn. Some of these meanings are compatible with my argument in this chapter, while others would require further qualifications. These qualifications, however, are not essential for my argument, which only involves locality in the sense just defined.

incompatible with the locality of quantum mechanics or of quantum phenomena. Hence, quantum mechanics is, thus far, as complete as nature appears to allow a theory of these phenomena to be. At least, according to Bohr, EPR's and Einstein's related arguments do not demonstrate otherwise.

At the same time, there exist interpretations of quantum mechanics, such as the one offered by Bohr in his reply, that are local. Accordingly, unless some arguments or some experimental data to the contrary become available, such interpretations allow one to see quantum mechanics as, within its scope, a complete and local theory of quantum phenomena. As Bohr's says in his reply: "a *viewpoint* termed 'complementarity' is explained *from which* quantum mechanical description of physical phenomena would seem to fulfill, within its scope, all rational demands of completeness" (Bohr 1935, p. 696; emphasis added).[3] Bohr's reply to EPR does not go quite as far as his later thinking and allows for an attribution of certain properties to quantum objects at the time of measurements under the constraints of the uncertainty relations, or even, with certain crucial qualifications (explained below), on the basis of predictions with certainty and hence independently of measurements.[4] At least, Bohr allows for the latter kind of attribution as part of his *counterargument* to EPR, since EPR grounds their argument in the assumption that we can do so. Bohr shows their argument is still insufficient even under this assumption. Bohr's *argument* itself in his reply virtually amounts to the impossibility of this assumption, in view of the qualifications just mentioned, which have especially to do with the question of locality. Bohr's aim is, again, also to offer an interpretation of quantum phenomena, those found in the EPR experiment included, and quantum mechanics as an adequate response to an "entirely new situation as regards the description of physical phenomena" (Bohr 1935, p. 700). In this interpretation, quantum mechanics appears as both a complete and a local theory of quantum phenomena, at the cost of not providing a space–time description of quantum processes.

From Einstein's viewpoint such a theory could not be considered complete, even if it is interpreted in terms of the "tamer" epistemology of Bohr's reply. It is not clear whether Einstein followed Bohr's subsequent arguments based on Bohr's ultimate epistemology, specifically that in "Discussion with Einstein," first published in "the Schilpp volume" (Schilpp 1949) and thus, at least ostensibly, on Einstein's radar by that time. In his comments on the EPR experiment in this volume, Einstein only responds to Bohr's argument in his reply to EPR, which he, I argue, misreads by attributing to Bohr an assumption of the nonlocality of quantum mechanics. However, it is difficult to think that Einstein would have found appealing a view that the actual

---

[3] Here and throughout his reply, Bohr uses the term phenomena in its conventional sense, rather than in his special sense, introduced later. Accordingly, in this chapter I shall use the term in this conventional sense as well, unless specified otherwise.

[4] Some commentators do attribute to Bohr, including in his post-EPR thinking, the view that "measurement reveals the objective, preexisting value of an observable" (Murdoch 1977, p. 107). This view is problematic in the present reading of Bohr, and I do not think that Murdoch offers sufficient support for his claim.

properties and behavior of the ultimate physical objects considered by a theory are inaccessible to the theory or inconceivable altogether. In all of his arguments of the EPR-type, Einstein assumes that quantum phenomena and quantum mechanics allow us to assign physical properties to quantum objects themselves on the basis of either measurements or predictions with certainty, possible in certain circumstances, such as those of the EPR-type experiments. In these experiments, these predictions are made on the basis of the data obtained from measurements performed on other quantum objects, spatially separated from those concerning which we make the predictions in question. Accordingly, it appears possible to assign these properties independently of any measurement performed on the objects themselves and hence "without in any way disturbing" them (Einstein et al. 1935, hereafter [*EPR* 1935], p. 138). On this basis, EPR and Einstein elsewhere argue that it is possible to ascertain that quantum objects independently possess *both* conjugate quantities, such as the position and the momentum of a quantum object. If correct, this argumentation would allow one de facto, although not in actual measurements, to circumvent the uncertainty relations. Since, however, it is never possible to ascribe definite values to both such quantities by means of the formalism of quantum mechanics, consistent with the uncertainty relations, quantum mechanics must be incomplete.

Bohr contested this argumentation by arguing that EPR do not show that it is possible to establish that a given quantum object can be assumed to simultaneously possess both conjugate quantities. Bohr bases his counterargument in reexamining the applicability to quantum phenomena of EPR's key assumptions, especially their criterion of physical reality: "*If, without in any way disturbing a system, we can predict with certainty (i.e., with probability equal to unity) the value of a physical quantity, then there exists an element of physical reality corresponding to this physical quantity*" (EPR 1935, p. 138). Bohr argues that quantum phenomena, as defined by the irreducible role of measuring instruments in their constitution, disallows the *way*—the *unqualified* and hence *ambiguous* way—this criterion is used by EPR, which disables the single most important part of EPR's logic.

Einstein thought that Bohr "came nearest to doing justice to the problem" posed by the EPR experiment, even though he did not think that Bohr had solved it. Einstein, however, appears to have misread Bohr's argument as allowing for a nonlocal character of quantum mechanics and preserving its completeness at this cost. Einstein did qualify that he "translated" Bohr's argument "into [his] own way of putting it" (Einstein 1949b, p. 681). Bohr's argument is lost in Einstein's "translation," which reads Einstein's own logic of the alternative between locality and completeness in quantum mechanics into Bohr's very different logic. Bohr's argument, based in "the singular position of measuring instruments in the account of quantum phenomena" (which makes it impossible to consider quantum objects apart from their interactions with these instruments), allows him to maintain both the completeness and locality of quantum mechanics.

Ironically, any classical-like complete theory that would predict these data now appears to have to be nonlocal in view of Bell's theorem and related findings. The situation would be similar to Bohmian hidden-variables quantum mechanics, where nonlocality is an explicit consequence of the formalism (in any version of it available so far). Bell's theorem tells us that any classical-like (hidden variables) theory, reproducing the statistical predictions of quantum mechanics, would be nonlocal, a finding further amplified by related theorems, such as the Kochen–Specker theorem, all of which, in the present view, appear to support Bohr's argument.[5] This argument did not go as far as rigorously proving the impossibility of local classical-like theories compatible with the data of quantum mechanics. Bohr did not demonstrate—in the way, for example, Bell's and the Kochen–Specker theorems do (for discrete variables)—that the data in question is logically incompatible with a joint assignment of both complementary properties to a given quantum object.[6] But he showed that, contrary to EPR's contention, quantum phenomena, including those of the EPR-type, disallow us to rigorously contemplate an experimental assignment of such properties simultaneously or even separately to *the same quantum object*. This demonstration disables the particular argument offered by EPR, although not all conceivable arguments against either completeness or locality of quantum mechanics. Bohr's argument does not guarantee that quantum phenomena cannot one day be proved to be nonlocal. But then relativity does guarantee this either. As things stand now, however, there is no experimental evidence that contradicts relativity, while there is much evidence to support it.

---

[5] There has been much debate concerning Bell's theorem and related findings: how tight these arguments are, what they actually demonstrate, and so forth. The subject has generated an immense body of literature, reflecting a great multitude of views, as have EPR's article and Bohr's reply, whether in their own right, or in the context of Bell's theorem, which renewed attentions to both papers. For some instructive exchanges concerning EPR's article and Bohr's reply, see (Mehra and Rechenberg 2001, vol 6, part 2, pp. 713–759). For reasonably representative selections of articles on Bell's theorem, see (Cushing and McMullen 1989), (Kafatos 1989), and (Ellis and Amati 2000). Bell's own work is assembled in (Bell 2004). See (Shimony 2004) and (Held 2006) for helpful introductions to, respectively, Bell's and the Kochen–Specker theorems. It is sometimes argued that Bell's theorem or the EPR-type experiments *imply* the nonlocality of quantum phenomena. I shall not try to counter these contentions here, although, in my view, Bell's theorem and these experiments only allow for the possibility of nonlocality, which, however, is in conflict with the experimental evidence, thus far confirming relativity. Apart from the fact that these contentions have no significant bearings on our understanding of Bohr's argumentation, they represent a minority view, albeit a vocal minority. My understanding of Bell's theorem, stated here, is consistent with that of many commentators. See, for example, Mermin's essays on the subject collected in (Mermin 1990) or his argument in (Mermin 1998), (Gottfried 2000), (Peres 1993), (Bertlmann and Zeilinger 2002), and (Zeilinger et al. 2005).

[6] Since they allow for attributing some (but not all) properties to a quantum object *at the time of measurement*, Bell's and the Kochen–Specker theorems only require the epistemology of the type Bohr adopts in his reply. They do not entail Bohr's ultimate epistemology, which is, however, consistent with both theorems.

8.2 "Can Quantum-Mechanical Description of Physical Reality Be Considered Complete?": EPR's Argument

EPR base their argument on two key criteria. One is a *necessary* criterion of completeness: for a theory to be considered complete, *"every element of the physical reality must have a counterpart in the physical theory"* (EPR 1935, p. 138). The criterion of completeness must be necessary for the type of argument they aim to make. On the other hand, with this criterion in hand, they only need a *sufficient* criterion of reality to demonstrate the incompleteness of quantum mechanics, at least under the condition of locality. EPR formulate their criterion of reality as follows: *"If, without in any way disturbing a system, we can predict with certainty (i.e., with probability equal to unity) the value of a physical quantity, then there exists an element of physical reality corresponding to this physical quantity"* (EPR 1935, p. 138). Although not always expressly stated, both criteria are retained in Einstein's subsequent arguments on the subject, where the center of his analysis shifts from the completeness of quantum mechanics to its locality. The completeness of quantum mechanics is still at stake in these arguments, since quantum mechanics is argued by Einstein to be either incomplete or, if complete, then nonlocal—the alternative considered in EPR's article as well.

EPR's argument for the incompleteness of quantum mechanics, by their criterion, depends on the assumption, taken by EPR to be self-evident, that this criterion applies as straightforwardly in quantum physics as it does in classical physics (EPR 1935, p. 139). This assumption becomes the main target of Bohr's counterargument. He argues that the criterion contains "an essential ambiguity … when it is applied to quantum phenomena" (Bohr 1935, p. 696). EPR's argument is, however, ingenious and, as Bohr noted later, "remarkable for its lucidity and apparently incontestable character," but, as he clearly implied, only *apparently* incontestable (Bohr 1949, *PWNB* 2, p. 59). I shall now outline the essential logic of this argument.

It may *appear* that the criterion of physical reality applies in the case of quantum phenomena and quantum mechanics as well, *without any further qualifications*, just as it does in classical mechanics. As EPR say: "Regarded not as a necessary, but merely as a sufficient, condition of reality, this criterion is in agreement with classical as well quantum-mechanical ideas of reality" (EPR 1935, p. 139). The reasons for this apparent or apparently unqualified applicability of the criterion to quantum phenomena are as follows. It is only a joint *simultaneous measurement* or *simultaneous prediction* of two conjugate quantities involved in the quantum-mechanical description that is impossible in view of the uncertainty relations. The value of a single variable can be measured with any degree of accuracy within the capacity of our measuring instruments. (As explained in Chap.6, the uncertainty relations are not affected by this capacity.) Hence in the idealized case this value can be considered precisely measurable, which fact is crucial to both EPR's argument and Bohr's reply.[7] The value of a single variable can also be *predicted exactly* without performing a measurement on a quantum object in certain idealized circumstances, such as those of the EPR experiment. The latter deals with two quantum objects, S_1 and S_2, forming an EPR pair (S_1, S_2), that have previously been in interaction but are then spatially separated. Once S_1 and S_2 are separated, one can establish (with certainty) *both* the *distance between* the two objects and *the sum of their momenta*. It can then be easily shown that, with the "states" (in the sense of formalism) of the two objects thus entangled and with these quantities in hand, by *measuring* either the position or, conversely, the momentum of S_1, one can *predict exactly* either the position or the momentum for S_2 without physically interfering with S_2.[8] This possibility is "an immediate consequence" of what Bohr calls "the transformation theorems of quantum mechanics," which reflect the fact that quantum mechanics offers "a coherent formalism covering automatically any procedure of measurement" (Bohr 1935, pp. 696–697n).

EPR thought that these facts enable one to argue for the incompleteness of quantum mechanics, or else, again, its nonlocality, which they see as impermissible. In Bohr's words, "According to their criterion, [EPR] therefore want to ascribe an element of reality [pertaining to S_2] to each of the quantities represented by such variables. Since, moreover, it is a well-known feature of the present formalism of quantum mechanics that it is never possible, in the description of the [physical] state of a mechanical system, to attach definite values to both of two canonically conjugate variables, they consequently deem this formalism to be incomplete, and express the belief that a more satisfactory theory can be developed" (Bohr 1935, p. 696). In other words, in EPR's view, quantum mechanics is a *correct* theory, insofar as it allows us to *predict* with certainty *some* among the physical quantities (elements of reality) associated with the phenomena in question. By the same token, it also enables us to ascertain, at least in situations of the EPR-type the existence *of all* such quantities in accordance with EPR's criterion of reality, or *so EPR and Einstein contend* (the contention contested by Bohr). However, quantum mechanics is incomplete by virtue of providing no means for *predicting* and, hence, determining *all of these quantities*, whose physical reality we are thus able to ascertain. EPR themselves sum up their argument and convey

---

[7] As I noted, unlike the Bell-Bohm version of the EPR experiment for spin, the original experiment proposed by EPR, dealing with continuous variables, cannot be physically realized, since the EPR entangled quantum state is non-normalizable. This fact, however, does not affect the fundamentals of the case, which can be considered in terms of the corresponding idealized experiment. Cf., Bohr's comment on this point (Bohr 1935, p. 698, n.). There are experiments that statistically approximate the idealized entangled state constructed by EPR.

[8] By "states" I mean here the mathematical entities (vectors in Hilbert spaces) that, accompanied by other elements of the formalism and appropriate rules, such as Born's rule, enable our predictions concerning the outcome of quantum experiments. One of the main questions in the debate concerning quantum mechanics is whether quantum states in this sense in fact correspond to some actual physical entities or "elements of physical reality." In Bohr's view, they do not.

their logic somewhat more cumbersomely (EPR 1935, p. 141). But their essential argument is, I think, properly conveyed by my summary, which follows that of Bohr in his reply. Before closing, EPR address the nonlocality of quantum mechanics as, they admit, a possible, but to them unacceptable, alternative, which I shall discuss in Sect. 8.4. Given that this alternative is unacceptable, EPR can now close their argument: "While we have thus shown that *the wave function does not provide a complete description of the physical reality,* we left open the *question* of whether or not such a description exists. We believe, however, that such a theory is possible" (*EPR* 1935, p. 141; emphasis added).

Essentially the same argument—that quantum mechanics is either incomplete or nonlocal, and that, given that nonlocality is not permissible, it must be seen as incomplete—remains in place in Einstein's subsequent writings on the subject. However dissatisfied Einstein might have been with EPR's article, he never renounced his claim that quantum mechanics is incomplete under the assumption of locality. He added the possibility that quantum mechanics is local as a statistical theory of ensembles, as opposed to individual quantum systems, which, however, he did not see as a way out of the dilemma, because a proper description of individual quantum systems would still be lacking. Accordingly, the argument of Bohr's reply remains applicable to these arguments by Einstein as well.

8.3 "Can Quantum-Mechanical Description of Physical Reality Be Considered Complete?": Bohr's Argument

Bohr's argument in his reply is based on the analysis of the irreducible (as against classical physics) role of measuring instruments in the constitution of quantum phenomena, a role that shaped Bohr's thinking concerning quantum mechanics and complementarity from the Como lecture on. Bohr argues that an examination of this role is lacking in EPR's article, even though EPR "emphasize" that "the extent to which an unambiguous meaning can be attributed to such an expression as 'physical reality' cannot of course … be deduced from *a priori* philosophical conceptions, but … must be founded on a direct appeal to experiments and measurements" (Bohr 1935, p. 696).

The main reasons why EPR's criterion of reality automatically applies in classical mechanics is that any object under investigation is presumed to possess the properties in question at any given point. The value of each such property may change in the course of the dynamic evolution of the object, and the equations of classical mechanics track these changes, thus also providing an idealized description of the systems considered. Classical mechanics makes predictions on the basis of this tracking. Whatever "disturbance" a measurement might introduce could, in principle, be neglected or compensated for. This also means that, for the purposes of physical description, one need not distinguish between objects and phenomena. The value of both conjugate variables could be established simultaneously and, in principle, within the same experimental arrangement, which ensures the causal and even, ideally, deterministic character of classical mechanics. Finally, it is also possible, ideally or in principle, to *exactly* repeat a given experiment on an *identically prepared* object or even on the same object.

These conditions no longer obtained in the case of quantum phenomena. As emphasized throughout this study, it is a well-established experimental fact that in this case we can control only the preparation of measuring instruments, because this preparation can be described classically, but never the outcomes of the corresponding experiments, which depend on quantum behavior, including the quantum interaction between quantum objects and measuring instruments. These outcomes are, in general, different, even when the apparatus is prepared identically, which makes the difference between quantum objects and quantum phenomena irreducible. It is, Bohr argues, by virtue of these circumstances, not sufficiently considered by EPR, that their application of their criterion of reality to quantum phenomena and quantum mechanics contains an essential ambiguity. It is possible to remove this ambiguity by qualifications that establish the proper conditions for using this criterion of reality in the case of quantum phenomena, including those defined by the EPR experiments. In effect, Bohr offers these qualifications. They concern the mutual exclusivity of the *two* measuring arrangements, always necessary for predicting alternative quantities in question, as against classical physics, where this can be done, at least in principle, in a single experimental arrangement. These qualifications, however, also make insufficient EPR's *argument* and, by implication, Einstein's later arguments concerning the subject. Once this ambiguity is removed by virtue of adding these qualifications, it can be shown that every element of reality that can be rigorously ascertained for a well-defined quantum phenomenon does have a counterpart in quantum mechanics, since the latter can predict this element. The probability of such a prediction is not always equal to unity, although sometimes it is, as in the EPR experiment. The number of elements of reality for a given quantum object thus ascertained could only be half of the number that EPR deem possible to ascertain. Thus, one could claim either an element of reality corresponding to a position measurement or an element of reality corresponding to a momentum measurement, but never both. The situation is, thus, strictly in agreement with the uncertainty relations, which are thus uncircumventable. As I have explained, EPR aim to de facto circumvent them by arguing that both variables could be *assigned* to a given quantum object, although both can never be *established simultaneously* by either a measurement or a prediction.

According to Bohr, this difference between quantum and classical phenomena arises by virtue of the following interrelated circumstances, which are also correlative to the irreducible probabilistic character of quantum predictions:

1. The first is *the necessity of discriminating in each case between quantum objects and measuring instruments*: "this necessity of discriminating in each experimental arrangement between those parts of the physical system considered which are to be treated as measuring instruments and those which constitute the objects under investigation may indeed be said to form a

principal distinction between classical and quantum-mechanical description of physical phenomena" (Bohr 1935, p. 701);
2. The second is *the irreducible role of measuring instruments in the constitution of quantum phenomena,* which reflects the necessity of discrimination stated in (1);
3. The third is, as a consequence, *the impossibility of rigorously considering the behavior of quantum objects independently of their interaction, their "finite [quantum] and uncontrollable interaction," with the measuring instruments* (Bohr 1935, pp. 697, 700).

It might appear that in the case of the EPR experiment the role of measuring instruments is no longer essential. This apparent possibility to dispense with this role was one of the main reasons why Einstein conceived of the experiment, in part in view of his previous exchanges with Bohr, where Bohr was able to use the role of measuring instruments to counter Einstein's criticism. As explained above, the EPR experiment involves an (EPR) *pair* of objects (S_1, S_2), which allows for predictions (with certainty) about S_2 without performing a measurement on it by performing a measurement on S_1, and hence without involving an interaction between S_2 and any measuring instrument which would "in any way disturb" S_2. Given this possibility, EPR argue *first,* on the basis of their criterion of reality, that the quantities thus *predicted* may be attributed to quantum objects themselves even at the time of the prediction (rather than at the time of a measurement), and hence that one need not disturb these objects by measurement in determining these quantities. *Second,* they argue that, since one can by means of such a procedure predict *either one or the other* of the two conjugate measurable quantities defining the behavior of physical systems (say, either the position or the momentum of the object), quantum objects independently possess *both* of these quantities, even though we can never predict both of them *simultaneously. Third* and finally, they argue that, because quantum mechanics provides no means for giving definite value to *both* such quantities in *the description of the same state of an object,* it must be incomplete. The underlined qualification is crucial, since one can, and EPR do, use quantum mechanics to predict each quantity in question separately, and Bohr will argue that, contrary to EPR's contention, we *cannot assign* both such quantities to the same state by any means, and hence do more than quantum mechanics does. The only alternative in EPR's view (but, as will be seen in Sect. 8.4, not that of Bohr) is that quantum mechanics is nonlocal, because assuming quantum mechanics to be complete, the state of a given quantum object could be determined by a measurement performed on another, spatially separated, quantum object.

Bohr argues that even in the case of the EPR experiment, the role of measuring instruments and the necessity of discriminating between them and quantum objects remain irreducible and entail limitations concerning the *types* of measuring arrangements used in determining one or the other quantity in question. These limitations apply even if this determination is done in terms of *prediction* rather than measurement in accordance with EPR's criterion of reality, and by means of performing this measurement on another quantum object. In Bohr's ultimate view, any attribution of properties to quantum objects at any time (even during a measurement) becomes ambiguous, including that of any single property, rather than only a joint attribution of complementarity properties. However, Bohr does not need this more radical epistemology for his counterargument to EPR. A "tamer" epistemological view, held by him at the time, is sufficient. This view allows for an assignment, in accordance with the uncertainty relations, of one and only one of the two conjugate measurable quantities—say, a position—to a given quantum object itself *at the time* and on the basis of a *measurement*, and only a measurement, rather than prediction, even with certainty, as in EPR's argument. This, Bohr argues, is not possible, or at least EPR's argument does not demonstrate this to be possible. Bohr's argument does contain or imply the provision that EPR's criterion is valid only if these predictions are *in principle verifiable*. This is not always the case in quantum physics, since an alternative measurement—say, that of momentum—on a quantum object for which a prediction regarding its position was made would erase the possibility of ever verifying this prediction. In classical physics this difficulty does not arise, since we can assign to the object both quantities simultaneously in all circumstances. This is one of the qualifications of EPR's criterion that quantum phenomena require and that is not considered by EPR. As will be seen below, this qualification is especially important for maintaining the locality of quantum phenomena and quantum mechanics.

EPR's main point is, again, that *both* conjugate physical quantities involved in the quantum-mechanical formalism can be ascribed *to the same quantum object,* albeit by means of two *separate* procedures, rather than simultaneously within the same procedure. This possibility, they argue, allows one to conclude that quantum mechanics is either incomplete, since it does not allow for such an assignment, or else nonlocal. Bohr's main *counter*-point, which disables this conclusion, is that even such a separate attribution of the second quantity in question *to the same quantum object* even by means of two separate procedures is never possible, once an attribution of the first one is made. This is, arguably, the single most crucial physical point behind—or, since Bohr does not expressly state it in this form, implied in—Bohr's argument in his reply, and it appears to be missed by most commentators on the exchange. Indeed, I am unaware of any commentary that addresses it, apart from the previous works by the present author. It is, however, essential in order to understand Bohr's thinking concerning the EPR experiment. EPR are not entirely unaware of this obstacle. They remark in closing that their criterion of reality might not be "sufficiently restrictive," and say that "one would not arrive at [their] conclusion if one insisted that two or more physical quantities can be regarded as simultaneous elements of reality *only when they can be simultaneously measured or predicted.* On this point of view, since either one or the other, but not both simultaneously, of the quantities P and Q can be predicted, they are not simultaneously real." However, they see this restriction as implying the nonlocality of quantum phenomena or of "reality" (EPR 1935, p. 141). As I argue here, the problem is rather the application of their criterion without qualifications that quantum phenomena *require* for this application, and as will be seen in the next section, these qualifications allow one to maintain locality as well.

EPR's and Bohr's arguments do share the same initial assumption, taken as axiomatic by each. Either in the standard or in the EPR case, one can setup a quantum-mechanical experiment in two alternative ways so as to predict *either* one *or* the other of the two conjugate measurable quantities associated with a given quantum object. In the EPR case, we can do so with probability equal to unity and without physically interfering with (disturbing) the object itself at the time of this prediction. I shall call this *"assumption A."*

From the possibility of the alternative covered by *assumption A,* EPR conclude that *both quantities can be assigned* to the same quantum object, even though it may not be possible to do so simultaneously, which may be designated as *"inference E"* (for Einstein). Accordingly, quantum mechanics is incomplete (unless one allows for nonlocality), and EPR express a hope that a future theory would enable us to offer a more complete description (EPR 1935, p. 141).

Bohr argues *inference E* to be impermissible, in view of the *conditions* under which any possible *association* of any measurable physical quantity with a quantum object can take place, most crucially the irreducible role of the measuring instruments in any such association. This role is irreducible even when such an *attribution* is made by means of predicting this quantity with certainty, in accordance with EPR's criterion of reality. Bohr assumes the possibility of such an attribution *for the purposes of his counterargument,* provided, again, that such a prediction is in principle verifiable. (As I said, in his own view, an attribution of a given single property to a quantum object itself is only possible on the basis of and at the time of measurement, and in his later view no such attribution is ever possible at all.) Bohr then argues that, given the irreducible role of measuring instruments, even if one assumes the EPR criterion of reality, a realization of the two alternative situations of measurement in question, which are necessary for the respective (verifiable) predictive assignment of these quantities, would *always* involve *two incompatible experimental arrangements.* It follows that, once one of these two quantities is thus established, it is never, in principle, possible to establish the second one; there is no conceivable experiment that would allow us to do so. For doing so would inevitably involve, in the standard case, *two different quantum objects* of the same type, such as electrons or photons, or in EPR's case, two EPR pairs, whose behavior, rather than only their initial preparation (defined by the same state of measuring instrument), will be *fully identical* in following the course of the experiment. This, however, is not possible. I shall designate this inference from *assumption A* as *inference B* (for Bohr).

The reason for this impossibility is, again, that quantum systems that are identically prepared in the sense of the state of measuring instruments involved do not, in general, behave identically. In fact, given that we can only control the classical physical state of (certain parts of) measuring instruments, it follows that the identical preparations of quantum systems themselves cannot be ascertained either, which is why Bohr refers to the interactions between quantum objects and measuring instruments as "finite [quantum] and uncontrollable" (Bohr 1935, pp. 697, 700). These circumstances entail the irreducibly probabilistic nature of all quantum predictions, regardless of the theory used to make them. This situation does not appear to be sufficiently taken into account by EPR, whose argument tacitly depends on the possibility of identically repeating the EPR experiment. However, whatever prediction one has made in an already performed EPR experiment, say, for the pair (S_{11}, S_{12}), one would require a *different (identically prepared) EPR pair* (S_{21}, S_{22}) for the measuring procedure, performed on S_{21}, necessary to make the alternative EPR prediction concerning S_{22}. This is also necessary if one wants to repeat the experiment in order to makes the same prediction, in which case the outcome can never be guaranteed to be the same and is usually different. This explains my double notations (S_{n1}, S_{n2}) in labeling the objects concerned. There is no physical situation in which one can ever assign both conjugate or complementary quantities to *the same object*—either simultaneously (the uncertainty relations) or separately in two possible alternative experiments— or to two fully identically behaving *objects* that will allow one, in principle, to justify the validity of such an assignment to the same object. Bohr does not make this last point expressly in this form. It is, however, a consequence of his argument (apparently lost on Einstein) that in the EPR experiment or related experiments contemplated by Einstein "we are not dealing with a *single* specified experimental arrangement, but are referring to *two* different, mutually exclusive arrangements" (Bohr 1949, *PWNB* 2, pp. 57, 60; Bohr 1935, p. 699).

One can diagrammatically represents the situation as follows (for the continuous variables, such as position and momentum, although one can easily extend the argument to discrete variables, such as spin). Let X and Y be two complementary variables in the Hilbert-space formalism $(XY - YX \neq 0)$ and x and y the corresponding physical measurable quantities $(\Delta x \Delta y \cong h)$; (S_1, S_2) is the EPR pair of quantum systems; and p is the probability of the prediction concerning a given variable for S_2, via the wave function Ψ on the basis of the measurement performed on the corresponding variable for S_1. The objects and quantities involved become doubled to (S_{11}, S_{12}) and (S_{21}, S_{22}), and correspondingly to (X_{11}, X_{12}), (X_{21}, X_{22}) and (Y_{11}, Y_{12}), (Y_{21}, Y_{22}) in the present view *(inference B)* of the situation, as against EPR's or Einstein's view *(inference E)*. Then:

The EPR experiment (EPR's and Einstein's view):

S_1		S_2
X_1	Ψ_1 (with $p = 1$) $\rightarrow$	X_2
Y_1	Ψ_2 (with $p = 1$) $\rightarrow$	Y_2

This view represents *inference E*, whereby both X_2 and Y_2 could be assigned to S_2, making quantum mechanics incomplete (or if complete, nonlocal).

The EPR experiment (Bohr's view):

S_{11}		S_{12}
X_{11}	Ψ_1 (with $p = 1$) $\rightarrow$	X_{12}
S_{21}		S_{22}
Y_{21}	Ψ_2 (with $p = 1$) $\rightarrow$	Y_{22}

This view represents *inference B,* which allows one to maintain both the completeness and the locality of quantum mechanics—or at least to argue that EPR did not prove otherwise, since their argument depends on *inference E,* which is not experimentally obtainable.

That this mutual exclusivity of the two experimental arrangements involved—specifically in the EPR experiment—cannot be avoided needs to be argued, and Bohr spends a large part of his reply doing so, basing his argument on "the finite and uncontrollable interactions between the objects and measuring instruments" (Bohr 1935, p. 700).[9] That such is the case should, however, be expected and the argument itself easily surmised from the discussion given in this study, especially the analysis of the double-slit experiment. Indeed, Bohr's argument amounts to a straightforward examination of the double-slit experiment, with only insubstantial additional considerations in the case of the EPR experiment. Bohr concludes in particular:

> [I]n the phenomena concerned we are not dealing with an incomplete description char-acterized by the arbitrary picking out of different elements of physical reality at the cost of sacrif[ic]ing other such elements, but with a rational discrimination between essentially different experimental arrangements and procedures which are suited either for an unambiguous use of the idea of space location, or for a legitimate application of the conservation theorem of momentum. Any remaining appearance of arbitrariness concerns merely our freedom of handling the measuring instruments, characteristic of the very idea of experiment. (Bohr 1935, p. 699)

This statement essentially reprises his definition of complementarity, instanti-ated in terms of the mutual exclusivity of certain experimental arrangements, while allowing us the freedom of selecting either one of them at any given point. The *physical* meaning of the uncertainty relations and, thus, the use of Heisenberg's formula itself, $\Delta q \Delta p \cong h$, are adjusted by Bohr in accordance with this view (cf., Bohr 1937, *PWNB* 4, p. 86).

The main point is that, given the mutual exclusivity of the measuring arrangements involved, the second quantity in question cannot even in principle ever be assigned to the *same quantum object, once one such quantity is assigned.* This holds true even if one accepts EPR's criterion of reality and allows that such an assignment is made on the basis of a prediction, because once an experiment enabling one to make the first prediction is performed, the first object, S_{11}, is no longer available to make the second prediction in question. A new EPR pair would be needed to "repeat" the experiment, *ab ovo,* to get to the stage of the prediction in question, and, as I shall explain presently, such a prediction can never be guaranteed to be properly coordinated with the first. For any given *single* quantum object, including the second object of the EPR pair, one can, in principle, predict or measure only *one* of the quantities involved—and *once one is predicted, never* the other. There is no conceivable experiment that would allow us to circumvent this doubling and its implications. This doubling is a manifestation of the

---

[9] For a detailed reading of Bohr's argument, see (Plotnitsky 2009, pp. 279–312).

irreducible individuality of all quantum phenomena, the *"individuality* completely foreign to classical physics" (Bohr 1935, p. 697). Each quantum event is each time unique, singular, and unrepeatable, which, again, makes quantum mechanics—or, it appears, any theory that could predict such events—irreducibly probabilistic even as concerns each event individually. Accordingly, while EPR's reasoning of considering, counterfactually, the alternative determinations of variables to the same object applies *in the case of a single classical object,* it cannot, contrary to EPR's contention, be even meaningfully contemplated *in the case of a single quantum object.*[10]

Nor is it possible (in the way it is in classical mechanics) to coordinate the two experiments in question on two different objects so as to make it possible to consider both as rigorously identically prepared *objects,* or EPR pairs. This is, again, because in dealing with quantum phenomena, we can only control our instruments in the same way, but never the behavior of the quantum objects and, consequently, the outcomes of the experiments defined by this behavior, which thus becomes subject to probabilistic estimates only. In the EPR case, we can predict with probability equal to unity the first quantity in question—say, the value of the position variable—for the second object, S_{12}, of a given EPR pair (S_{11}, S_{12}). We can then predict the second quantity—the value of the momentum variable— for the second object, S_{22}, of, unavoidably, *another, "identically prepared,"* EPR pair (S_{21}, S_{22}). However, we cannot coordinate these predictions in such a way that they could be considered as pertaining to two identically prepared *objects* in the way this could be done in classical physics. This is not possible since the necessary intermediate measurements would, in general, give us different data. Were we to repeat the measurement and the prediction of the first pair of quantities, those of the position variables for respectively S_{21} and S_{22}, we could still make our pre- diction with the probability unity, but the outcome that we predict will, in general, not be the same as in the case of the first pair (S_{11}, S_{12}). In other words, we *can predict* the outcome of a given EPR experiment with probability equal to unity but we *cannot repeat* such an experiment with probability equal to unity for the values of the corresponding outcomes. We can only coordinate such measurements and predictions statistically and thus establish the EPR correlations, a fact used by Bell in the case of discrete (spin) variables and Bohm's version of the EPR experi- ment.[11] This does not help EPR, since their argument—either for the

---

[10] These considerations bear on the question of counterfactual statements in considering quantum phenomena. This question, also germane to Bell's and the Kochen–Specker theorems, is beyond my scope here. See the works cited in Note 5, especially (Mermin 1990) and (Peres 1993).

[11] The argument given here could be transferred to Bohm's version of the EPR experiment and spin variables. In this case, too, any assignment of the alternative spin-related quantity to the *same* quantum object becomes impossible, once one such quantity is assigned. An assignment of the other would require an alternative type of measurement, mutually exclusive with the first, on the first object of a given pair, and hence another fully identically behaving EPR-Bohm pair, which is, again, not possible. Only statistical correlations between such assignments are possible

incompleteness or, alternatively, for the nonlocality of quantum mechanics—de facto presupposes exact, rather than statistical, coordination of such variables as belonging to the same (the only case EPR themselves consider), or an *identically prepared* object of the same or an identically prepared EPR pair. Quantum mechanics properly predicts these outcomes, and hence it corresponds to what obtains experimentally, while, as this argument shows, what EPR argue for *(inference E)* does not obtain experimentally.

Bohr's statement that, in view of his analysis, there is an essential ambiguity in EPR's use of their criterion, and specifically that this ambiguity concerns EPR's expression "without in any way disturbing the system,'" has caused much confusion (Bohr 1935, pp. 696, 700). This statement occurs in the most famous elaboration of Bohr's reply, and arguably most misunderstood even by sympathetic commentators, primarily because Bohr's overall argumentation is not adequately considered in addressing the elaboration. I would like, accordingly, to offer a reading of this elaboration in relation to Bohr's argument, as explained above.[12] I shall divide the elaboration into key units, because it is, as will be seen presently, crucial for an understanding of Bohr's point about how one parcels and groups these units, and I shall suggest a grouping different from that used in most readings. Bohr says:

[1] From our point of view we now see that the wording of the above-mentioned criterion of physical reality proposed by Einstein, Podolsky, and Rosen contains an ambiguity as regards the meaning of the expression "without in any way disturbing a system."

[2] Of course there is in a case like that just considered no question of a mechanical disturbance of the system under investigation during the last critical stage of the measuring procedure.

[3] But even at this stage there is essentially the question of *an influence on the very conditions which define the possible types of predictions regarding the future behavior of the system.*

[4] Since these conditions constitute an inherent element of the description of any phenomenon to which the term "physical reality" can be properly attached, we see that the argumentation of the mentioned authors does not justify their conclusion that quantum-mechanical description is essentially incomplete.

[5] On the contrary this description, as appears from the preceding discussion, may be characterized as a rational utilization of all possibilities of unambiguous interpretation of measurements, compatible with the finite [quantum], and uncontrollable interaction between the object and the measuring instruments in the field of quantum theory.

[6] In fact, it is only the mutual exclusion of any two experimental procedures, permitting the unambiguous definition of complementary physical quantities, which provides room for new physical laws the coexistence of which might at first sight appear irreconcilable with the basic principles of science. It is just this entirely new situation as regards the description of physical phenomena that the notion of *complementarity* aims at characterizing. (Bohr 1935, p. 700; Bohr's emphasis)

(Footnote 11 continued)
(cf., Mermin 1990, pp. 107–108). The argument concerning locality given in the next section could be transferred to the case of discrete variables as well.

[12] Ideally, one would need a proper reading of Bohr's discussion leading to this elaboration, which cannot be offered here, but, see, again (Plotnitsky 2009, pp. 279–312). However, the logic of Bohr's argument, explained above is sufficient to explicate Bohr's main meaning.

The invocation of "ambiguity" in [1] and Bohr's claim that this ambiguity arises "as regards the meaning of the expression 'without in any way disturbing a system'" is often taken in conjunction with Bohr's claim [3]. This reading would imply that, according to Bohr, while there is no "mechanical disturbance" of the second, spatially separated, object, S_2, of the EPR pair (S_1, S_2), by performing any measurement upon it, there is (contrary to EPR's claim) some form of "disturbance" of or "influence" on S_2 arising from our manipulations of S_1 on S_2, thus implying some form of nonlocal influence. As must be apparent from the analysis given above, it is difficult to argue that this is what Bohr has in mind here. There appears to be little, if anything, to support such a view either in [3] or in the overall elaboration in question, or in Bohr's argument in his reply. Instead, this ambiguity has to do with the conditions under which—in the EPR situation or in the case of quantum phenomena in general—an unambiguous meaning can be assigned to the terms, such as EPR's "elements of reality," that define the situation in question. These conditions are defined by the irreducible nature of the interactions between quantum objects and measuring instruments. As explained above, once one conjugate quantity in question is established (even on the basis of a prediction, in accordance with EPR's criterion of reality, via the corresponding measurement on S_{11}) for S_{12}, we cannot ever establish the second quantity involved without performing the corresponding measurement upon it and, in this sense, but only in this sense, *without disturbing* it. We can only establish such a quantity without disturbing the object for a different quantum object, S_{22}, of a different EPR pair (S_{21}, S_{22}), by a different measurement on S_{21}. Thus, the ambiguity in question clearly relates to the clause "without in any way disturbing the system," insofar as EPR claims that both quantities could be established for the one and the same object, the same second object of a given EPP pair, without disturbing this object. This clause requires a qualification if one wants to apply it unambiguously in the EPR situation, since, as explained here, we can never determine both conjugate quantities for the second object of any EPR pair without disturbing this object, but only one of these quantities. Once one conjugate quantity in question is established (even on the basis of a prediction, in accordance with EPR's criterion of reality) for S_{12}, we cannot ever establish the second quantity involved without measuring and hence *disturbing* S_{12}. We can establish such a quantity only for a different quantum object, S_{22}, via a different EPR pair (S_{21}, S_{22}), by a different measurement on S_{21}. Nor, as noted above, can these two determinations be coordinated so as to assume that both quantities could be associated with the same object of the same EPR pair. Bohr, again, *agrees* with EPR that we cannot allow for any "disturbance," physical, or other, of anything that is spatially separated from actual physical measurement.

Bohr's meaning turns on how one reads Bohr's "But" in [3], because it can be linked to two "sequences," *Sequence A* and *Sequence B*. *Sequence A* is from [2]: "Of course there is in a case like that just considered no question of a mechanical disturbance of the system under investigation during the last critical stage of the measuring procedure" to [3]: "But even at this stage there is essentially the question of *an influence on the very conditions which define the possible types of predictions regarding the future behavior of the system." Sequence B* is from [2]: "Of course there is in a case like that just considered no question of a mechanical disturbance of the system under investigation during the last critical stage of the measuring procedure," thus taking locality as an axiom, to [3] *and* [4]: "But even at this stage there is essentially the question of *an influence on the very conditions* [physically having to do with our manipulation of the first object of a given EPR pair] *which define the possible types of predictions regarding the future behavior of the system* [the second object of this pair]," *and* "since these conditions constitute an inherent element of the description of any phenomenon to which the term 'physical reality' can be properly attached [unambiguously applied], we see that the argumentation of the mentioned authors does not justify their conclusion that quantum-mechanical description is essentially incomplete." That is, [3] and [4] must be considered together! Important, too, is Bohr's point that the influence in question is on the conditions defining possible types of *predictions* concerning the second object and not on the behavior of this object. These conditions have to do with our manipulation and disturbance of the *first* object of the pair, and *not the second*.

Thus, an ambiguity as concerns the expression "without in any way disturbing the system" does not arise because there is, contrary to EPR's claim, a physical (mechanical) disturbance or influence between the two spatially separate physical situations in question. On this point, Bohr agrees with EPR. Instead, it is a question of our manipulations of—our "influence" on—S_1. Bohr refers to his argument that, given the mutually exclusive conditions of measurements and predictions in EPR's case, the two alternative predictions in question can be made only in two mutually incompatible experimental setups and, it follows, only for two different quantum objects. Neither these conditions nor this inescapable fact are specified by EPR, thus making their criterion ambiguous and their argument, based on the view that this can be done for the *same* quantum object, unsuitable for demonstrating either the incompleteness or, as will be seen, the nonlocality of quantum mechanics. It is, thus, the *reality* of this situation that, while it is coherently covered by quantum mechanics, it stands in the way of EPR's *criterion* of reality, which need not necessarily correspond to nature, and thus in the way of their argument. This is why Bohr says that the EPR criterion of physical reality contains "an [essential] ambiguity as regards the *meaning* of the expression 'without in any way disturbing the system' " (Bohr 1935, p. 700). It is an obvious consequence of Bohr's interpretation that in any quantum-mechanical measurement, the language of "disturbance" cannot apply in the sense of assuming that there are any undisturbed properties of quantum objects that are then disturbed by measurement.

It is thus clear that the "influence" in Bohr's statement does not refer to any physical influence of our interference with S_1 upon S_2 or any system, such as a measuring device, locally associated with S_2 (concerning which we make predictions but upon which we do not perform measurements). Instead, it refers to the physical influence upon—a *fixing* of—the measurement-prediction situation defined by the particular setup of measuring instruments physically associated with and only with S_1 (upon which we do perform measurements and with which we, hence, interfere). As a result, this influence determines—it "influences"—what kind of *predictions* we can or cannot make concerning the second object. It defines the conditions of one or the other of the two—always irreducibly mutually exclusive—situations in the EPR-type experiments or indeed in all quantum-mechanical predictions. If the apparatus is setup for measuring one complementary quantity for the measuring system associated with the first particle, then within this setup the other variable cannot be *unambiguously defined*—we are absolutely precluded from doing so—for either S_1 or S_2. That is, such is the case, again, unless we independently perform an alternative measurement on S_2—which, however, amounts to a change of setup and disables any possible determination of the initial variable or quantity for it. The influence Bohr has in mind is clearly that upon the conditions of an unambiguous definition of the quantities in question, and in this sense, "influence" may not be the best term to use here. These conditions physically affect only S_1, where direct physical measurements are performed, although the prediction and the possibility of definition themselves also concern S_2, which is not physically influenced, and nor is anything that can be associated with it. This is why, while Bohr says that there is *"an influence on the very conditions which define the possible types of prediction regarding the future behavior of the system,"* he never says that anything disturbs, interferes with, or even influences S_2 itself: "there is … no question of a mechanical disturbance of the system under investigation." This last statement is clearly made with EPR's alternative between the locality and completeness of quantum mechanics in mind, which I shall consider in detail in the next section.

In sum, Bohr questions the unambiguous applicability of EPR's criterion of reality and the adequacy of their argumentation, given the nature of the phenomena under consideration, while, as follows from the nature of the "influence" in question, remaining in agreement with the locality requirements, since there is no physical influence here. Accordingly, and given that quantum mechanics properly responds to the situation, EPR's conclusion that it is incomplete need not follow. Hence Bohr concludes:

> "[5] On the contrary this description, as appears from the preceding discussion, may be characterized as a rational utilization of all possibilities of unambiguous interpretation of measurements, compatible with the finite and uncontrollable interaction between the objects and the measuring instruments in the field of quantum theory."
>
> [6] In fact, it is only the mutual exclusion of any two experimental procedures, permitting the unambiguous definition of complementary physical quantities, which provides room for new physical laws the coexistence of which might at first sight appear irreconcilable with the basic principles of science. It is just this entirely new situation as regards the description of physical phenomena that the notion of *complementarity* aims at characterizing. (Bohr 1935, p. 700)

The concept of complementarity allows one to avoid the kind of ambiguity inherent EPR's criterion reality, unless this criterion is properly qualified, which complementarity essentially does. In the process, it also provides a proper scientific grounding of the physical laws—encoded in the formalism of quantum mechanics—that otherwise appear "irreconcilable with the basic principles of science." Of course, it does so at the cost of suspending certain other principles, which some, specifically Einstein, see as equally basic, such as the possibility of offering a realist and preferably causal description of quantum objects and processes, at least, by way of idealized models. However, even if nature *may* allow for such a theory, EPR do not demonstrate that nature actually *does so*. Nor, given the character of the influence in question, does it follow, as an alternative, that quantum mechanics is nonlocal, as EPR contend.

8.4 Can Quantum-Mechanical Description of Physical Reality Be Considered Local?

In closing their article, EPR acknowledge that they did not demonstrate that one could ever *simultaneously* ascertain by measurement or prediction both quantities in question for the same quantum object, such as S_1 or S_2 in their experiment. However, they see this qualification as implying nonlocality in the EPR situation and hence as unreasonable. According to them:

> One could object to this conclusion [that the quantum-mechanical description of physical reality given by wave functions is incomplete] on the grounds that our criterion of reality is not sufficiently restrictive. Indeed, one would not arrive at our conclusion if one insisted that two or more physical quantities can be regarded as simultaneous elements of reality *only when they can be simultaneously measured or predicted.* On this point of view, since either one or the other, but not both simultaneously, of the quantities P and Q can be predicted, they are not simultaneously real. This makes [in the EPR case] the reality of P or Q depend upon the process of measurement carried out on the first system, which does not disturb the second system in any way. No reasonable definition of reality could be expected to permit this. (EPR 1935, p. 141)

In Einstein's subsequent arguments this alternative serves as a starting point, from which he proceeds to argue, along the lines of EPR's argument, that, since nonlocality is unacceptable, quantum mechanics cannot be considered complete by his criteria, roughly of the EPR-type.

Now, nonlocality does follow, if, along with the completeness of quantum mechanics, one assumes, as EPR do, that the measurement, say, of P, on S_1 *fixes the physical state* or reality of S_2 by "a spooky *action* at a distance," rather than allows for what may be called "a spooky *prediction* at a distance" by fixing *the possible conditions* of such a prediction, as discussed above. It is true that in this latter view, taken by Bohr, quantum mechanics becomes a strictly predictive, and moreover, only probabilistically predictive, theory concerning our possible knowledge regarding a future state of a given system, or even only of a measuring apparatus impacted by that system. Our EPR predictions are "spooky" because there is no explanation how the outcomes we predict come about. The nature of these outcomes cannot be explained, which, again, need not assume that they come about because of physical action at a distance. The question is, again, whether nature allows us to do more than offer such predictions. EPR reason that, if quantum mechanics is assumed to be complete, a given measurement—say, that of position, Q, on S_1—and the wave function, cum Born's rule, give us as much knowledge as possible concerning the physical state of S_2, in this case leaving the momentum for S_2 completely unknown. This is correct (leaving aside for the moment, but only for the moment, the question whether this knowledge concerns the *actual physical state* of S_2 or is the best knowledge we can have concerning its *possible state*). Since, however, we can also perform an alternative measurement, that of momentum, P, and use the wave function and Born's rule to predict the momentum for S_2, now with the position of S_2 left completely unknown, the state of S_2 is different. Accordingly, the state of S_2—established, again, without in any way disturbing it—depends on the measurement performed on the spatially separated system S_1. Hence, quantum mechanics is, *if complete*, nonlocal. Under EPR's assumption, an alternative measurement—say, in the first case (when Q is predicted), of P on S_2 —would discontinuously change this fixed state. Although EPR do not examine this last eventuality, it is an automatic consequence of their argumentation, as Einstein realized later (e.g., Einstein 1949a, p. 81).

Their key point, however, is, again, that the physical reality of S_2 depends, as it does under this assumption, on whether we measure P or Q of S_1, and it is differently determined accordingly. This entails nonlocality, assuming that quantum mechanics is complete. Or, as Einstein, again, argued later, one is left with a paradoxical situation insofar as (assuming that quantum mechanics is complete) two mutually incompatible states could be assigned to the same quantum system. The paradox, Einstein said, is resolved by assuming that the state of S_2, as defined by the wave function, is determined by a measurement performed on a spatially separated object, S_1, or, more strongly, "by denying altogether that spatially separated entities posses independent real states," either of which assumptions he saw as "entirely unacceptable" (Einstein 1949a, p. 81). Yet another alternative, discussed in the next chapter, would be an assumption that quantum mechanics is a theory of ensembles, rather than individual systems, which, however, would still make the theory incomplete in Einstein's view.

Einstein thought that Bohr accepted the alternative of locality versus completeness, and retained completeness by allowing for nonlocality, which would also imply that Bohr accepted that part of EPR's argument that was based on their initial less restrictive criterion of reality (Einstein 1949b, p. 681). However, as I mentioned earlier, Einstein misread Bohr, who, again, only allows for a spooky *prediction*, and *not action*, at a distance. This is why Bohr argues in his reply— clearly with EPR's alternative between locality and completeness in mind—that it is not a question of a physical or, as he says, "mechanical" influence of the measurement on S_1 upon the physical state of affairs concerning S_2. Hence, he sees quantum mechanics as local. As he expressly stated in his reply, "the singular position of measuring instruments in the account of quantum phenomena ... together with the relativistic invariance of the uncertainty relations ... ensures the compatibility between [the] argument [of his reply] and all exigencies of relativity theory" (Bohr 1935, p. 701n.; also Bohr 1938, *PWNB* 4, pp. 105–106; Bohr 1958, *PWNB* 3, p. 3).

In Bohr's view, physical states cannot be seen as properly determined (even when we have predicted them exactly) unless either the actual measurement is made, or there is a possibility of actually *verifying* the prediction. In other words, it must be possible to perform an actual measurement that would yield the predicted value. This condition in turn becomes a necessary qualification of EPR's criterion of reality in the case of quantum phenomena. Hence, even though it is impossible ever to *make* both predictions in question for the same quantum object, the question of the *verification* of an EPR's prediction, or any quantum-mechanical prediction, remains important. For, if one assumes the validity of EPR's criterion in its original (unrestrictive) form, the measurement of the alternative quantity, Q (the coordinate), on S_2 would automatically disable any possible verification of the original prediction concerning P (the momentum), by virtue of the mutual exclusivity of the two measuring arrangements involved. In other words, once this alternative measurement is performed, the original prediction becomes meaningless as in principle unverifiable; it is *erased* for any meaningful purposes—for example, for any further predictions concerning either object, S_1 or S_2. This erasure is also correlative to the impossibility of repeating an identically prepared experiment with the same or a properly correlated outcome. An object with a known momentum may be a legitimate concept in quantum physics, just as it is in classical physics, but it is meaningful only insofar as a given measurement enables (as Bohr puts it in his reply to EPR) a corresponding "correlation between its behavior and some instrument" (Bohr 1935, pp. 699–700). This correlation itself is found in classical physics as well. The difference is that measuring a conjugate quantity, that of the object's position, still allows us, at least in principle, to measure the momentum simultaneously—and hence to ascertain both qualities simultaneously. This is never possible in the case of quantum phenomena, because an alternative measurement always erases our previous predictions altogether, making irrelevant the information that enabled us to make those predictions.

This erasure of quantum predictions by an alternative measurement appears to have been yet another reason why Bohr was ultimately compelled to see any measurable quantities in question in quantum theory as determined only through measurement and never through prediction, and as pertaining only to the (classical) physical aspects of measuring instruments involved or to phenomena in his new sense. The impossibility of ever verifying an actual spatial–temporal attribution of even a single property to quantum objects themselves (as opposed to measuring instruments) and even at the time of measurement (rather than only independently) appears to have been another key factor. On both counts, however, the possibility of, in principle, verifying a given prediction appears to be crucial to Bohr's thinking.

Einstein's subsequent arguments on the subject are sometimes viewed as offering a stronger case by focusing more sharply on the question of nonlocality (e.g., Einstein 1936, 1948, in Born 2005, pp. 166–170, 204–205, 210–211; Einstein 1949a, pp.77–85; and, in commenting on Bohr's reply, Einstein 1949b, pp. 681–682). These arguments do streamline EPR's argument, in part by this shift of focus. One the other hand, they still make their case in terms of the alternative between locality and completeness along the lines of EPR's argument. In particular, they never consider—any more than EPR's argument does—that, once the first of the two conjugate variables for the second, distant quantum object of an EPR pair is predicted (using a measurement on the first one), the second conjugate variable can never, in principle, be predicted for the same object. Some of these arguments add discussions of quantum mechanics as a theory of ensembles (which resolve the problem of nonlocality), but as will be discussed in Chap. 9, this possibility still leaves quantum mechanics incomplete, according to Einstein.

It is true that neither in his reply to EPR nor elsewhere in his discussions of the EPR experiment—such as (in terms of his concept of phenomenon) in the Warsaw lecture (1938) or "Discussion with Einstein" (1949)—does Bohr *directly* invoke the subject of locality, qua locality, as perhaps he should have done. Part of the reason appears to be that, in EPR's article in particular, it is the question of completeness that is the main focus of Einstein's argument, and this question is raised in all of Einstein's arguments, in contrast to more recent discussions (following Bell's theorem), which usually assume quantum mechanics to be complete and center solely on locality. Bohr, especially in his reply to EPR, responds primarily to this question of the completeness of quantum mechanics, which appeared to have posed greater concerns at the time. It is clear, however, that Bohr is well aware of the role of locality in quantum theory, and, as I explained earlier in this chapter, he thought on the subject along the EPR lines as early as 1925. Some of his exchanges with Einstein involve Bohr's arguments based on relativity, and his writings often address the question of the relative epistemological positions of relativity and quantum mechanics, or of both vis-à-vis classical physics. As I argue here, Bohr took the requirement of relativity and hence locality as axiomatic, just as Einstein did, and Bohr expressly noted in his reply "the compatibility between [his] argument and *all exigencies* of relativity theory" (Bohr 1935, p. 701n.; emphasis added). The locality of quantum mechanics, he argues, is assured by the irreducible role of the measuring instruments, along with the relativistic invariance of the uncertainty relations, and is not undermined by EPR's argument (Bohr 1935, p. 701, 701n.).

8.5   Postscript. 1927–1949. "Discussion with Einstein on Epistemological Problems in Atomic Physics" (A Post-EPR View)

Bohr was not unprepared to respond to EPR by his previous thinking concerning quantum phenomena, especially by his previous exchanges with Einstein, which I would like to discuss in this postscript. While some of these exchanges took place earlier and thus preceded EPR's paper, Bohr's recount of these exchanges, given primarily in his 1949 "Discussion with Einstein," is clearly inflected by his exchange with EPR and by his ultimate epistemology, to be discussed in the next chapter. It is, however, the logic of Bohr's argument in his reply (only minimally affected by his ultimate epistemology) that is most essential for understanding Bohr's argument is these exchanges. Accordingly, it is fitting to consider them here.

The experiments discussed in these exchanges show that the conditions of the EPR-*type*—and hence the argument offered here—apply to *all* quantum predictions, although there are some inessential differences between the standard and the EPR case. The reasons for this applicability are as follows. In any quantum measurement, we actually *predict* the value of either the position or the momentum associated with a quantum object *after* the object has already left the region of its interaction with the measurement apparatus, in which interaction defines the data on the basis of which we make the prediction (Bohr 1949, *PWNB* 2, p. 57). Hence, we make this prediction (at a distance) without *any further* interference with the quantum object under investigation, whether we assume that this quantity is already defined by this prediction, in accordance with EPR's criterion, or that it is yet to be defined by a subsequent measurement, in accordance with Bohr's ultimate view (Bohr 1949, *PWNB* 2, p. 57).

In the EPR situation, which involves two quantum objects rather than a single object and a measuring apparatus, the case is a bit more complicated, but it is not fundamentally different (Bohr 1935, pp. 699–700). The fact of the previous interaction between the objects involved, S_1 and S_2, brings the EPR and the standard case closer together. A certain quantum part of the instruments involved appears in the same role as the first object, S_1, of an EPR pair (S_1, S_2), which has previously interacted with the second object, S_2. The registered measured data relates to this part of the instrument in a way similar to the data obtained by performing the final measurement (there are certain intermediate stages) on S_1. By the same token, a prediction we make on the basis of this data will concern an object (the one under investigation) separated from this part of the instrument. Conversely, the first object, S_1, may be treated as a quantum part of a measuring instrument, although one would still need to add a proper classical part to it, which is essentially done by performing a measurement on it in any event, since this measurement involves a measuring apparatus. In this case, as against that of the standard measurement, we are no longer dealing with predictions concerning S_1 but only with those concerning S_2, which is what makes the analogy in question possible. Bohr came to realize this analogy at least by the time of the Warsaw lecture (Bohr 1938, *PWNB* 4, pp. 101–103; Bohr 1949, *PWNB* 2, p. 60). Of course, in the standard case, as opposed to that of the EPR experiment, a prediction in question—concerning, say, the location of a collision between the object and the silver bromide screen—will not, in general, have a probability equal to unity, but some other probability. In order to get a probability equal to unity, one would need to perform intermediate measurements, just as one would in the EPR case. Such measurements are, in general, not possible in the case of the standard quantum measurement, since the quantum part of the apparatus, with which the object under investigation interacts, cannot be defined to create a traceable EPR situation. It is only possible to treat the corresponding part of the apparatus as a quantum object and reproduce the EPR case by adding another measuring apparatus, say, for measuring the behavior of a diaphragm in the double-slit arrangement. As Bohr explains: "If [in the double-slit experiment arrangement], for the two parts of the [EPR-type] system, we take a particle and a diaphragm, ... we see that the possibilities of specifying the state of the particle by measurements on the diaphragm just corresponds to the situation ... [in which] after the particles has passed the diaphragm, we have in principle the choice of measuring either the position of the diaphragm or its momentum [by a corresponding instrument in each case], and, in each case, making predictions as to subsequent observations pertaining to the particle. As repeatedly stressed, the principle point here is that such measurements demand mutually exclusive experimental arrangements" (Bohr 1949, *PWNB* 2, p. 60). Hence, again, they cannot be performed on the same quantum objects.

These differences, however, do not affect the main point at the moment, which is as follows. All our predictions concerning quantum objects, that is, concerning their future interactions with measuring instruments, are defined by their previous interactions with other quantum objects, whether a part of the quantum strata of measuring instruments or one of the objects of the EPR pair. Moreover, as Bohr explained (anticipating Wheeler's delayed choice experiment), we can always make our plans and setup one arrangement or another after this interaction has already taken place (Bohr 1935, pp. 698–699; Bohr 1949, *PWNB* 2, p. 57). The situation thus becomes more strictly parallel to that of the EPR case, in which we decide, after the interactions between S_1 and S_2 took place, whether to measure the position or, conversely, the momentum associated with S_1 and hence which prediction to make concerning S_2, and make the corresponding, mutually exclusive, experimental arrangements. In either case, once this determination has been made, any possibility of making an alternative prediction concerning the other conjugate variable associated with this second object, or again, any quantum object concerning which we make a prediction, is unavoidably precluded. Accordingly it is, in principle, impossible to speak of an alternative determination concerning the same object. Such alternative determinations could only concern two different quantum objects, just as in the EPR experiment.

Bohr's account of the first set of these exchanges during the Solvay conference in Brussels in October 1927 in "Discussion with Einstein" suggests the beginning of the development of Bohr's thinking in the direction just outlined, as applied to the double-slit experiment, on which he bases his reply to EPR as well. The language of "effects" and "phenomena" reflects his post-EPR thinking and inflects the account itself, as throughout "Discussion with Einstein." Bohr says:

> This point [of the double-slit experiment] is of great logical consequence, since it is only the circumstance that we are presented with a choice of *either* tracing the path of a particle *or* observing interference effects, which allows us to escape from the paradoxical necessity of concluding that the behaviour of an electron or a photon should depend on the presence of a slit in a diaphragm through which it could be proved not to pass. We have here to do with a typical example of how the complementary phenomena appear under mutually exclusive experimental arrangements ... and are just faced with the impossibility, in the analysis of quantum effects, of drawing any sharp separation between an independent

behavior of atomic objects and their interaction with the measuring instruments which serve to define the conditions under which the phenomena occur. (Bohr 1949, *PWNB* 2, pp. 46–47)

Of course, in order to observe interference effects we would need many repeated trials, but, as explained earlier, our probabilistic prediction for any single particle would be affected by what kind of arrangement we make as well. The situation acquires further features relevant to the EPR experiment in the discussion of the famous photon box experiment proposed by Einstein and subsequent (refined) arguments by Einstein, eventually leading him to his arguments of the EPR-type. What makes Einstein's new argument especially significant is the shift from the question of a possible *measurement* of both conjugate variables to that of a possible *prediction* of both. Originally, the argument was formulated as follows, according to Bohr:

> Einstein proposed the device … consisting of a box with a hole in its side, which could be opened or closed by a shutter moved by means of a clock-work within the box. If, in the beginning, the box contained a certain amount of radiation and the clock was set to open the shutter for a very short interval at a chosen time, it could be achieved that a single photon was released through the hole at a moment known with as great accuracy as desired. Moreover, it would apparently also be possible, by weighing the whole box before and after this event, to measure the energy of the photon with any accuracy wanted, in definite contradiction to the reciprocal indeterminacy of time and energy quantities in quantum mechanics. (Bohr 1949, *PWNB* 2, p. 53)

Einstein apparently accepted Bohr's counterargument showing, by using Einstein's own general relativity, that one cannot really circumvent the uncertainty relations in this way (*PWNB* 2, pp. 53–55). The tightness of this counterargument has been questioned sometimes, but this is not that important at the moment. Bohr's counterargument is crucial for the development of Bohr's thought, both in its own right and as part of the sequence it forms with the next stages of the exchange, proceeding via experiments of the EPR-type and discussed earlier in this chapter. Accordingly, of principal interest here is Bohr's essential logic of differentiating between those parts of measuring instruments and (quantum) objects under investigation and his conclusion. He summarizes this logic as follows: "The discussion, so illustrative of the power and consistency of relativistic arguments, thus emphasizes once more the necessity of distinguishing, in the study of atomic phenomena, between the proper measuring instruments which serve to define the reference frame and those parts which are to be regarded as objects under investigation and in the account of which quantum effects cannot be disregarded" (Bohr 1949, *PWNB* 2, pp. 55–56). His conclusion is that, in accordance with the (time–energy) uncertainty relations (again, seen as experimentally given), "a use of the apparatus as a means of accurately measuring the energy of the photon will prevent us from controlling the moment of its escape" (Bohr 1949, *PWNB* 2, p. 55). Two mutually exclusive arrangements would be required for an exact measurement of each variable, with an obvious implication that the same logic would apply to any pair of complementary variables. Hence, again, an assignment of such quantities would require at least two *identically* behaving quantum objects, and there is no experimental arrangement that could ever guarantee such a behavior in identically prepared experiments. This fact disables considering the variables in question as pertaining to the same objects.

Anticipating EPR's argument and criterion of reality, Einstein's follow-up argument shifts the questioning toward *predictions* rather than measurements, which, in Bohr's words, "might seem to enhance the paradoxes beyond the possibilities of logical solution." According to Bohr:

> Thus, Einstein had pointed out that, after a preliminary weighing of the box with the clock and the subsequent escape of the photon, one was still left with the choice of either repeating the weighing or opening the box and comparing the reading of the clock with the standard time scale. Consequently, we are at this stage still free to choose whether we want to draw conclusions either about the energy of the photon or about the moment when it left the box. Without in any way interfering with the photon between its escape and its later interaction with other suitable measuring instruments, we are, thus, able to make accurate predictions pertaining *either* to the moment of its arrival *or* to the amount of energy liberated by its absorption. Since, however, according to the quantum-mechanical formalism, the specification of the state of an isolated particle cannot involve both a well-defined connection with the time scale and an accurate fixation of energy, it might thus appear as if this formalism did not offer the means of an adequate description. (Bohr1949, *PWNB* 2, p. 56)

Bohr also observes that "paradoxes of the kind contemplated by Einstein are encountered also in such simple arrangements [as those used in the double-slit type of experiments]. In fact, after a preliminary measurement of the momentum of the diaphragm, we are in principle offered the choice, when an electron or a photon has passed through the slit, either to repeat the momentum measurements or to control the position of the diaphragm and, thus, to make predictions pertaining to alternative subsequent observations" (Bohr 1949, *PWNB* 2, p. 57). He adds a comment that anticipates Wheeler's delayed choice experiment: "it obviously can make no difference, as regards observable effects obtainable by a definite experimental arrangement, whether our plans of constructing or handling the instruments are fixed beforehand or whether we prefer to postpone the completion of our planning until a later moment when the particle is already on its way from one instrument to another" (Bohr 1949, *PWNB* 2, p. 57).

While noting that "once more Einstein's searching spirit has elicited a peculiar aspect of the situation in quantum theory, which in a most striking manner illustrated how far we have here transcended customary explanation of natural phenomena," Bohr could not agree with "the trend" of Einstein's argument (Bohr 1949, *PWNB* 2, pp. 56–57, 59). The reasons for Bohr's problems with this "trend" should be clear in view of the preceding discussion. The experiments in question would always have involved two mutually exclusive arrangements and, hence, two different quantum objects, again, just as they do in the EPR case (where we need two different EPR pairs of quantum objects). Accordingly, Einstein's argumentation does not adequately correspond to the situation that obtains in quantum phenomena. As Bohr says:

[W]e must realize that in the problem in question we are not dealing with a *single* specified experimental arrangement, but are referring to *two* different, mutually exclusive arrangements. In the one, the balance together with another piece of apparatus like a spectrometer is used for the study of the energy transfer by a photon; in the other, a shutter regulated by a standardized clock together with another apparatus of similar kind, accurately timed relative to the clock, is used for the study of the time of propagation of a photon over a given distance. In both these cases, as also assumed by Einstein, the observable effects are expected to be in complete conformity with the predictions of the theory.... The problem again emphasizes the necessity of considering the *whole* experimental arrangement, the specification of which is imperative in any well-defined application of the quantum-mechanical formalism. (Bohr 1949, *PWNB* 2, p. 57)

Self-evidently, these two mutually exclusive measurements cannot apply to the same quantum object. Accordingly, the arguments, both that of Einstein and that of Bohr, are analogous to those of the EPR case, as Bohr says: "[W]e are here [in the EPR case] dealing with problems of just the same kind as those raised by Einstein in previous discussions" (Bohr 1949, *PWNB* 2, p. 59).

It is, again, true that in the EPR-type experiments the predictions in question are enabled by the measurements performed on a different—rather than, as in the argument just considered, the same—quantum object, which has previously been in interaction with the object under investigation. This is what allows EPR to speak, in accordance with their criterion, of predictions "without in any way disturbing the system" and, moreover, predictions with certainty, in an apparent contrast with the standard situation. Nevertheless, *essentially*, the EPR situation is analogous to the standard one. The standard predictions are always made concerning the object—analogous to the second object, S_2, of a given EPR pair—which, at the time of the prediction, is spatially separated from the measuring devices, whose quantum stratum plays the role analogous to that of the first object, S_1, of the pair, upon which we perform a given EPR measurement. Hence, Bohr's argument applies in both situations, and in his reply he considers a particular (double-slit) measuring arrangement for the EPR experiment, which makes this parallel especially apparent. There are, again, differences concerning the probabilities involved, but they do not affect the essential points in question. Although ingenious, the EPR contrivance of making such predictions on the basis of performing a measurement on a different quantum object so as not to interfere with the object in question does not change the essential aspects of the situation and hence does not help Einstein's case for the incompleteness, or else nonlocality, of quantum mechanics.

Chapter 9
1937–1938. "Complementarity and Causality" and "The Causality Problem in Atomic Physics" (The Warsaw Lecture): The Knowable and the Unthinkable

9.1 New Concepts and New Epistemology

This chapter considers Bohr's ultimate interpretation of quantum phenomena and quantum mechanics in terms of (in addition to complementarity) his concepts of phenomena and atomicity, defined by the effects of the interactions between quantum objects and measuring instruments, effects observed in these instruments. At the same time, quantum objects themselves and their behavior are placed beyond not only description but also any conception we can form. This view of the situation also entails a particular understanding of probability, to be discussed in this chapter as well. I proceed as follows. After a brief introduction, offered in this section, Sect. 9.2 considers the ultimate version of Bohr's interpretation, as defined by the concepts of phenomena and atomicity. Section 9.3 is a discussion of quantum probability from the perspective of this interpretation. It includes a discussion of Einstein's arguments concerning the incompleteness of quantum mechanics as, in his view, only a statistical theory of ensembles rather than a theory describing individual quantum processes. I conclude with a postscript on possible applications of Bohr's concept of complementary and his epistemology beyond physics.

Bohr's ultimate interpretation of quantum phenomena and quantum mechanics emerged following his exchange with Einstein concerning the EPR-type arguments, as discussed in the preceding chapter. This interpretation was not yet in place in his reply to EPR, which essentially used the interpretation developed by Bohr around 1928–1929, considered in Chap. 5. It is difficult to say when exactly this new version was finalized. It appears that it happened somewhere in the late 1930s, and, I would surmise, shortly after Einstein's important 1936 article "Physics and Reality" (Einstein 1936), published about a year before this interpretation first transpires in Bohr's articles. As discussed in Chap. 8, "Physics and Reality" was expressly mentioned by Bohr in a related context in "Discussion with Einstein" (Bohr 1949, PWNB 2, pp. 62–63). This interpretation was certainly in place by the time of "Discussion with Einstein," published in 1949 and

A. Plotnitsky, *Niels Bohr and Complementarity*, SpringerBriefs in Physics,
DOI: 10.1007/978-1-4614-4517-3_9, © Arkady Plotnitsky 2013

sometimes seen as Bohr's most definitive exposition of his views in their final form. The article, however, does not appear to present new developments of Bohr's thinking or to have been aimed to do so. It was conceived as an overview of Bohr's "discussion with Einstein," written in honor of Einstein, and was published in "the Schilpp volume" (Schilpp 1949). It may be added that, although the article traces the *chronology* of these exchanges, it does not reflect on the development and, especially, changes in Bohr's thinking, in part under the impact of these exchanges, as discussed in this book. Instead, the article presents Bohr's earlier argument through the optics of his ultimate views. As noted earlier, Bohr rarely, if ever, reflects on these changes and tends to present his interpretation, from the Como lecture on, as relatively stable, at most, indicating certain refinements of his presentation of his argumentation (e.g., Bohr 1935, p. 696; Bohr 1949, PWNB 2, p. 61).

I would argue that the two articles, published in quick succession, "Causality and Complementarity" in 1937 (originally given as a lecture in 1936) and "The Causality Problem in Atomic Physics," published in 1938 and also known as the Warsaw lecture (it was given in Warsaw in the same year), are where Bohr's new interpretation first appeared in print. Both closely follow the appearance of Einstein's "Physics and Reality" in 1936. While "Complementarity and Causality" contains the idea of the irreducibly inaccessible nature of quantum objects and their behavior (without quite speaking in these terms), the concepts of phenomena and atomicity are introduced in the Warsaw lecture. The latter also contains crucial reflections on probability, likely in part in response to Einstein's argument, presented in "Physics and Reality," concerning the statistical character of quantum mechanics as, in his view, an incomplete theory of quantum phenomena. The appearance of the term "causality" in the titles of both articles is not coincidental, although the term is found in other Bohr's titles and reflects Bohr's persistent concern with the question of causality.

9.2 Phenomena, Atomicity, and Quantum Objects

As we have seen in this study, Bohr's thinking and his interpretation of quantum phenomena and quantum mechanics, and correlatively, his instantiations of the concept of complementarity had undergone refinements and sometimes revisions. Throughout these developments, however, from the Como lecture on, this thinking has always been based on his understanding of the constitutive, irreducible role of measuring instruments in defining quantum phenomena. Eventually, in the wake of his exchange with EPR, Bohr was compelled to introduce the concept of phenomenon, defined by the effects of the interactions between quantum objects and measuring instruments, effects classically manifest in measuring instruments. In Bohr's words:

I advocated the application of the word *phenomenon* exclusively to refer to the observations obtained under specified circumstances, including an account of the whole experimental arrangement. In such terminology, the observational problem is free of any special intricacy since, in actual experiments, all observations are expressed by unambiguous statements referring, for instance, to the registration of the point at which an electron arrives at a photographic plate. Moreover, speaking in such a way is just suited to emphasize that the appropriate physical interpretation of the symbolic quantummechanical formalism amounts only to predictions, of determinate or statistical character, pertaining to individual phenomena appearing under conditions defined by classical physical concepts. (Bohr 1949, PWNB 2, p. 64; emphasis added)

One might prefer to speak of *an* (rather than *the*) appropriate interpretation. The key features of Bohr's concept are, however, apparent here—in particular, the point that the term refers "to the observations obtained under specified circumstances" and hence to already *registered* phenomena. Bohr's concept, thus, includes a rigorous specification of each arrangement, determined by the type of measurement or prediction we want to make, which specification also reflects the irreducibly individual, unique character of each phenomenon. Thus, if seen independently of the quantum-mechanical context of its appearance, each mark on the screen in the double-slit experiment, as discussed in Chap. 6, would perceptually or phenomenally appear the same regardless of the difference in the physical conditions and, hence, outcome (interference" or "no interference) of the double-slit experiment. According to Bohr's understanding, however, each mark is, or is part of, a *different individual* (quantum) phenomenon depending on these conditions, which are mutually exclusive in the case of complementary phenomena and are defined by each phenomenon uniquely in any circumstances. It follows that, in the double-slit experiment, rather than dealing only with two phenomena, each defined by a different multiplicity of spots on the screen, we deal with two distinct multiplicities of individual phenomena. Each spot on the screen is an individual phenomenon in Bohr's sense and depends on a different set of conditions of the experiment, although the statistical considerations pertinent in the experiment (including the fact that a given mark cannot be guaranteed to correspond to an emission from the source) still apply. One of these sets will manifest the interference pattern. The other will not.

Individually, in the first case, we deal with the momentum measurement (occurring at the diaphragm with the slits) and, in the second, the position measurement (occurring at the slit), associated with each event and, physically, pertaining to the measuring apparatus. None of these measurements is sufficient to determine the course of the future event and where the object hits the screen exactly (or what will be the object's momentum at a given later time), but only the probability is that it will do so within a certain region (or will have a momentum within a certain range at given time). While, thus, a given single event does not allow us to establish in which setting it had occurred, one can still see quantum mechanics as a probabilistic theory of individual quantum events. The main point at the moment is that any quantum phenomenon is always unique and unrepeatable, and as such is incompatible with any other actual situation of measurement. These individual situations may not all be as significant, especially for Bohr, as are complementary situations, which are correlative to the uncertainty relations. The irreducible individuality of each quantum phenomenon is, however, *prior* to complementarity and, thus, more important foundationally. What arguably matters most is this individuality and the probabilistic nature of our predictions concerning each individual phenomenon. That some of these phenomena are complementary or subject to the uncertainty relations is almost a secondary matter, as Feynman noted (Feynman 1985, p. 55n3).

Far from being a matter of convenience, the distinction between two multiple-spot phenomena and two multiplicities of different individual spot-like phenomena in the double-slit experiment is essential. The statistical qualifications do not diminish this point, but instead reinforce it. Given this type of view, first, as discussed in Chap. 6, no paradoxical properties—such as simultaneous possession of contradictory wave-like and particle-like attributes on the part of quantum objects—are involved, which allows one to dispense with the wave-particle complementarity. Second, we should never mix considerations that belong to complementary experimental setups in analyzing a given experimental outcome even when dealing with a single spot on the screen, as we could, in principle, do in classical physics (e.g., Bohr 1938, PWNB 4, p. 103). This is not an uncommon problem, including in some of Einstein's arguments.

We always have a free choice as concerns what kind of experiment we want to perform, in accordance with the very idea of experiment, which defines classical physics as well (Bohr 1935, p. 699). Unlike in classical physics, however, implementing our decision concerning what we want to do will allow us to make only a certain type of prediction (for example, that concerning a future position measurement) and will unavoidably exclude the possibility of certain other, complementary types of prediction (in this case, that concerning a future momentum measurement). Moreover, our freedom only allows us to select and control the initial setup of any given experiment but not its outcome, which fact also ensures the objective verifiability or falsifiability (objectivity) of our experiment and the unambiguous and unambiguously communicable character of our statements. The "objectivity" in this sense becomes an important emphasis in Bohr's writing (e.g., Bohr 1954, PWNB 2, p. 67; Bohr 1958, PWNB 3, p. 7). Bohr summarizes the overall argument just outlined in his 1954 "Unity of Knowledge":

A most conspicuous characteristic of atomic physics is the novel relationship between phenomena observed under experimental conditions demanding different elementary concepts for their description. Indeed, however contrasting such experiences might appear when attempting to picture a course of atomic processes on classical lines, they have to be considered as complementary in the sense that they represent equally essential knowledge about atomic systems and together exhaust this knowledge. The notion of complementarity does in no way involve a departure from our position as detached observers of nature, but must be regarded as the logical expression of our situation as regards objective description in this field of experience. The recognition that the interaction between the measuring tools and the physical systems under investigation constitutes an integral part of quantum phenomena has not only revealed an unsuspected limitation of the mechanical conception of nature, as characterized by attribution of separate properties to physical systems, but has forced us, in the ordering of experience, to pay proper attention to the conditions of observation. (Bohr 1954, PWNB 2, p. 74; also p. 73)

The appeal to "detached observers" brings to mind Einstein and Pauli, as representing two contrasting views of the situation, and Bohr might have had both in mind as well. For Einstein, the quantum-mechanical observer is not sufficiently *detached* and, hence, not sufficiently objective.[1] In Bohr's view of the situation, however, the observers are as detached vis-à-vis measuring instruments as they are in classical physics, thus ensuring the objectivity of the scheme. On the other hand, the measuring instruments involved can never be "detached" from quantum objects, because these objects cannot be extracted from the closed observed phenomena, containing them (Bohr 1954, PWNB 2, p. 73). Phenomena cannot be opened so as to reach quantum objects themselves by disregarding the role of measuring instruments in the way it is possible in classical physics or relativity, and thus in conflict with Einstein's desiderata for fundamental physics. For Pauli, as noted in Chap. 7, the quantum-mechanical observer is still too detached, at least for a successful approach to quantum field theory, in view of the fact that the theory does not take into account the atomic structure of measuring instruments.

The concept of phenomena allowed Bohr to resolve some of the difficulties and ambiguities of his own previous arguments and to bring greater clarity to his conception and presentation of the epistemology of quantum theory. As he also noted in "Discussion with Einstein," right before his passage outlining his concept of phenomenon, cited above: "I warned especially against phrases, often found in the physical literature, such as 'disturbing of phenomena by observation' or 'creating physical attributes to atomic objects by measurements.' Such phrases, which may serve to remind [us?] of the apparent paradoxes in quantum theory, are at the same time apt to cause confusion, since words like 'phenomena' and 'observation,' just as 'attributes' and 'measurement,' are used in a way hardly compatible with common language and practical definition" (Bohr 1949, PWNB 2, pp. 63–64; also Bohr 1938, PWNB 4, p. 104; Bohr 1954, PWNB 2, p. 73). It is difficult to fault Bohr on any aspects of this statement—*as this statement relates to his view at the time, his post-EPR view of the situation.* Given the preceding discussion in this study, however, it is also clear that Bohr's own earlier arguments sometimes suffer from insufficient clarifications in this respect, which his concept of phenomenon helped to remedy.

A crucial part of Bohr's concept of phenomenon is that it precludes any description or even conception of quantum objects themselves and their behavior, which behavior is, nevertheless, responsible for the emergence of these phenomena and the effects defining each such phenomenon. Physical quantities obtained in quantum measurements, such as those defining the physical behavior of certain (classically described) parts of measuring instruments, can no longer be assumed to represent the corresponding properties of quantum objects, even any single such

---

[1] Curiously, in "Discussion with Einstein," Bohr invokes Einstein's "*detached* attitude," which left "a deep impression" on him (Bohr 1949, PWNB 2, p. 36; emphasis added).

property, rather than only certain joint properties, in accordance with the uncertainty relations. As we have seen, Bohr's earlier view allows this type of attribution at the time of measurement. Even this weaker view, however, implies that the physical state of an object cannot be defined on the model of classical physics, which requires an unambiguous determination of both conjugate quantities for a given object at any point and independently of measurement. In Bohr's ultimate view, an attribution *even of a single property* to any quantum object as such is *never possible—before, during, or after measurement.*[2]

Bohr's reasons for moving to this argumentation are not difficult to perceive. The conditions that experimentally obtain in quantum experiments allow us to rigorously specify measurable quantities that can only, in principle, be physically associated strictly with measuring instruments and never with quantum objects themselves. Even when we do not want to know the momentum or energy of a given quantum object and thus need not worry about the uncertainty relations, neither the exact *position* of this object itself nor the actual time at which this "position" is established is ever available and hence in any way verifiable. These properties, assuming they could be defined (as they can be in some interpretations of quantum mechanics or in Bohmian theories), are lost in, to return to the language of Bohr's reply to EPR, "the finite [quantum] and uncontrollable interaction" between quantum objects and measuring instruments. However, this process leaves a mark in measuring instruments, which can be treated as a part of a permanent, objective record. Such records we *can* think and speak about, communicate, and so forth. This view appears to be expressed for the first time in "Complementarity and Causality":

> [T]he whole situation in atomic physics *deprives of all meaning such inherent attributes as the idealizations of classical physics would ascribe to the object...* The renunciation of the ideal of causality in atomic physics which has been forced upon us is founded logically only *on our not being any longer in a position to speak of the autonomous behavior of a physical object,* due to the unavoidable interaction between the object and the measuring instruments which in principle cannot be taken into account, if these instruments according to their purpose shall allow the unambiguous use of the concepts necessary for the description of experience. In the last resort an artificial word like "complementarity" which does not belong to our daily concepts serves only briefly to remind us of the

---

[2] This view may also help us to understand Bohr's famous (reported) statement, apparently in response to the question "whether the algorithm of quantum mechanics could be considered as somehow mirroring an underlying quantum world": "There is no quantum world. There is only an abstract quantum physical description. It is wrong to think that the task of physics is to find out how nature is. Physics concerns what we can say about nature" (reported in Petersen 1985, p. 305). One must, again, exercise extreme caution in considering such extemporaneous comments, especially when they are reported. There are, however, supporting statements clearly made by Bohr. With Bohr's ultimate epistemology, as here considered, in mind, this statement may be best read not as denying the existence of quantum objects but as denying the applicability of any conceivable description or even conception to them and their behavior, including "quantum" and "world," or "object" and "behavior." Beginning with Heisenberg's invention of it, as discussed in Chap. 3, "an abstract quantum physical description" (the quantum-mechanical formalism) only provides a set of algorithms for predicting, probabilistically, the outcomes of quantum experiments.

epistemological situation here encountered, which at least in physics is of an entirely novel character. (Bohr 1937, PWNB 4, pp. 86–87; emphasis added)

There is no qualification here or in Bohr subsequent writings to the effect that such "inherent attributes" may still be meaningful if their assignment is constrained by the uncertainty relations (e.g., Bohr 1949, PWNB 2, pp. 40, 51, 61). The uncertainty relations remain valid, of course, but they now only apply to the corresponding (classical) variables of suitably prepared measuring instruments, impacted by their interactions with quantum objects. We can either prepare our instruments so as to measure a change of momentum of certain parts of those instruments or arrange them so as to locate a spot impacted by a quantum object, but never do both together. In "Discussion with Einstein," Bohr expressed this view of the uncertainty relations as follows:

[A]n adequate tool for a complementary way of description is offered precisely by the quantum-mechanical formalism which represents a purely symbolic scheme permitting only predictions, on lines of the correspondence principle, as to results obtainable under conditions specified by classical concepts. It must be remembered that even in the indeterminacy relations [$\Delta q \Delta p \cong h$] we are dealing with an implication of the formalism which defies unambiguous expression in words suited to describe classical physical pictures. Thus, a sentence like "we cannot know both the momentum and the position of an atomic object" raises at once questions as to the physical reality of two such attributes of the object, which can be answered only by referring to the conditions for the unambiguous use of space–time concepts, on the one hand, and dynamical conservation laws, on the other hand. While the combination of these concepts into a single picture of a causal chain of events is the essence of classical mechanics, room for regularities beyond the grasp of such a description is just afforded by the circumstance that the study of the complementary phenomena demands mutually exclusive experimental arrangements. (Bohr 1949, PWNB 2, pp. 40–41)

The commentary ostensibly refers to the Como lecture but is clearly inflected by Bohr's post-EPR thinking and language, in particular his argument concerning "an essential element of ambiguity ... [always] involved in ascribing conventional physical attributes to atomic objects," again, even single such attributes (Bohr 1949, PWNB 2, p. 40). "Unity of Knowledge" expressly describes the uncertainty relations via the concept of phenomenon:

In particular, the impossibility of a separate control of the interaction between the atomic objects and the instruments indispensable for the definition of the experimental conditions prevents the unrestricted combination of space–time coordination and dynamical conservation laws on which the deterministic description in classical physics rests. In fact, any unambiguous use of concepts of space and time refers to an experimental arrangement involving a transfer of momentum and energy, uncontrollable in principle, to fixed scales and synchronized clocks which are required for the definition of the reference frame. Conversely, the account of phenomena which are characterized by the laws of conservation of momentum and energy involves in principle a renunciation of detailed space–time coordination. These circumstances find quantitative expression in Heisenberg's indeterminacy relations which specify the reciprocal latitude for the fixation of kinematical and dynamical variables in the definition of the state of a physical system. In accordance with the character of the quantum mechanical formalism, such relations cannot, however, be interpreted in terms of attributes of objects referring to classical pictures, but we are here dealing with the mutually exclusive conditions for unambiguous use of the very concepts of space and time on the one hand, and of dynamical conservation laws on the other. (Bohr 1954, PWNB 2, pp. 72–73)

Indeed, as explained in Chaps. 6 and 8, both variables involved in the uncertainty relations cannot be seen as pertaining to the same quantum object, which circumstance also makes the uncertainty relations correlative to the probabilistic nature of our quantum predictions and gives them an essentially statistical physical meaning.

Bohr's concept of phenomenon reflects the fact that observable quantum phenomena (in the usual sense) are defined by certain effects of the interactions between quantum objects and measuring instruments, "practically irreversible amplification effects," such as a click of a photo-detector or a blackening of a grain of a photographic emulsion (Bohr 1949, PWNB 2, p. 51). The very language of effects becomes persistent in Bohr's post-EPR writings. Thus, this concept of phenomena is the culmination of Bohr's argument for "the *impossibility of any sharp separation between the behavior of atomic objects and the interaction with the measuring instruments which serve to define the conditions under which the phenomena appear*" (Bohr 1949, PWNB 2, pp. 39–40). It is, again, crucial that the concept of phenomenon only refers to already *registered* observations, in other words to what has already happened and *not* to what may happen, even if on the bases of a rigorous prediction, which one expects to be confirmed by an experiment.

Bohr's insistence on the indispensability of classical physical concepts in considering the measuring instruments is a subtle and often misunderstood issue. Most especially, as I have stressed earlier, the possibility of classical description only applies to certain strata of measuring instruments, while certain other strata of their constitution, those that are responsible for their interactions with quantum objects, are seen by Bohr as quantum. Thus, on the one hand, "although, of course, the existence of the quantum of action is ultimately responsible for the properties of the materials of which the measuring instruments are built and on which the functioning of the recording devices depends, this circumstance is not relevant for the problems of the adequacy and completeness of the quantum-mechanical description in its aspects here discussed" (Bohr 1949, PWNB 2, p. 51; also Bohr 1937, PWNB 4, p. 88).[3] On the other hand, the dependence itself of the constitution and functioning of the recording devices on the quantum of action is crucial, since it enables the quantum interaction between these devices and quantum objects, without which there would be no quantum data. Earlier in the same paragraph, Bohr speaks of "the purely classical *account*" of measuring apparatuses involved (Bohr 1949, PWNB 2, p. 51; emphasis added). In other words, the account we use in processing the registered results of measurements is that of classical physics, but the overall understanding of measuring apparatuses is both classical, at that end, and quantum, at the other end, where they interact with quantum objects. No physical description of what

---

[3] As Bohr notes on the second occasion cited here and as was discussed in Chap. 7, the situation changes once we move to quantum field theory, where the quantum constitution of measuring instruments might need to be taken into account.

happens in these interactions as such is possible, anything more than a physical description of quantum objects themselves, concerning which, that is, again, concerning the effects of their future interactions with measuring instruments, our predictions are made. According to Bohr:

> [W]e must recognize above all that, even when the phenomena transcend the scope of classical physical theories, the account of the experimental arrangements and the recording of observations must be given in plain language, suitably supplemented by technical physical terminology. This is a clear logical demand, since the very word "experiment" refers to a situation where we can tell others what we have done and what we have learned. However, the fundamental difference with respect to the analysis of phenomena in classical and quantum physics is that in the former the interaction between the objects and the measuring instruments may be neglected or compensated for, while in the latter this interaction forms an integral part of the phenomena. The essential wholeness of a proper quantum phenomenon finds indeed logical expression in the circumstance that any attempt at its well-defined subdivision would require a change in the experimental arrangement incompatible with the appearance of the phenomenon itself.
>
> ... every atomic phenomenon is closed in the sense that its observation is based on registrations obtained by means of suitable amplification devices with irreversible functioning such as, for example, permanent marks on a photographic plate, caused by the penetration of electrons into the emulsion. In this connection, it is important to realize that the quantum-mechanical formalism permits a well-defined application referring only to such closed phenomena. (Bohr 1954, PWNB 2, pp. 72–73; also Bohr 1949, PWNB 2, p. 51)[4]

A registered quantum event is, thus, defined, as in classical physics, by, as Bohr noted in his reply to EPR, a "correlation" between the object under investigation and the measuring apparatus involved (Bohr 1935, pp. 669–700). Unlike in classical physics, however, this "correlation" is only manifested in the corresponding effect and the closed phenomenon associated with this effect, while "ascribing customary physical attributes to atomic objects," again, even single attributes (rather than only conjugate ones in view of the uncertainty relations), or any conceivable attributes becomes ambiguous, and is ultimately precluded (Bohr 1949, PWNB 2, p. 51).

The wholeness or indivisibility of phenomena in Bohr's sense is defined by all the features just outlined, most especially, again, by the irreducible distinction between the classically describable effects of the interactions between quantum objects and measuring instruments and quantum objects, which are indescribable by quantum theory or any means available to us, and ultimately are inconceivable. The irreducible nature of this distinction is not inconsistent with the wholeness of phenomena, quite the contrary, because the latter results from the fact that

---

[4] Bohr's concept of "irreversible amplification" may be related to "decoherence" approaches to quantum theory (Zurek 2003; Schlosshauer 2007). There is, however, a crucial difference. In Bohr's view, at least after the Como lecture, a quantum interaction between a quantum object and a measuring instrument only leaves an "irreversibly amplified" classical trace, which, or rather the corresponding numerical data, can be (probabilistically) predicted by quantum mechanics. But there is no *physical quantum state* that is described by the formalism and that "decoheres" into this trace, as in most decoherence approaches, such as that of Zurek.

quantum objects cannot be observed independently and, hence, cannot be seen as identical with objects as in classical mechanics. Technically, phenomena and objects are also different in classical mechanics, but this difference can be disregarded for the purposes of the idealized physical description. It cannot be in quantum physics, which prevents the use of idealized descriptive models of the classical type. Consider, this important elaboration from Bohr's reply to EPR, reflecting the beginning of his thinking, eventually leading him to his concept of phenomena and his ultimate view of the situation, as just outlined:

> [The] necessity of discriminating in each experimental arrangement between those parts of the physical system considered which are to be treated as measuring instruments and those which constitute the objects under investigation may indeed be said to form a *principal distinction between classical and quantum-mechanical description of physical phenomena.* It is true that the place within each measuring procedure where this discrimination in made is in both cases largely a matter of convenience. While, however, in classical physics the distinction between objects and measuring agencies does not entail any difference in the character of the description of the phenomena concerned, its fundamental importance in quantum theory, as we have seen, has its root in the indispensable use of classical concepts in the interpretation of all proper measurements, even though the classical theories do not suffice in accounting for the new types of regularities with which we are concerned in atomic physics. In accordance with this situation there can be no question of any unambiguous interpretation of quantum measurement other than that embodied in the well-known rules [e.g., Born's rule] which allow to predict the results to be obtained by a given experimental arrangement described in a totally classical way, and which have found their general expression through the transformation theorems, already referred to [as implying the possibility of the EPR predictions]. By securing its proper correspondence with the classical theory, these theorems exclude in particular any imaginable inconsistency in the quantum mechanical description, connected with a change of place where the discrimination is made between object[s?] and measuring agencies. In fact, it is an obvious consequence of the above argumentation [that given in Bohr's reply to EPR] that in each experimental and measuring procedure we have only a free choice of this place within a region where the quantum-mechanical description of the process concerned is effectively equivalent with the classical description. (Bohr 1935, p. 701)

The distinction between the two domains, where each description, quantum and classical, would respectively apply is sometimes known as the cut, the term more commonly used by Heisenberg and von Neumann than by Bohr, and known as the Heisenberg or Heisenberg-von Neumann cut (von Neumann 1932, pp. 418–420). Heisenberg based his unpublished commentary of EPR's paper on the examination of the cut and the possibility, noted by Bohr here, of shifting the place where the cut is made (Heisenberg 1935). Heisenberg's commentary was known to Bohr and possibly influenced this elaboration, more germane to Bohr's general epistemology of quantum phenomena than to his (rather than Heisenberg's) response to EPR's argument, which is why I consider it here, rather than in my discussion of Bohr's reply to EPR. That "a given experimental arrangement described in a totally classical way," again, only refers to the relevant observable parts of measuring instruments, and not to their ultimate constitution, especially as concerned those parts of this constitution that are involved in the quantum interactions between quantum objects and measuring instruments. It should also be noted that Bohr's statement does not imply that, while parts of measuring instruments are described by means of classical physics, the independent behavior of quantum objects is described by means of the quantum-mechanical formalism. As discussed in Chap. 4, while this view of the situation is not uncommon, it is not that of Bohr, after he revised the Como argument. Bohr obviously does say that *parts* of measuring instruments are described by classical physical concepts, again, keeping in mind that this description only concerns parts of measuring instruments and that, moreover, "the classical theories do not suffice in accounting" for the observed physical behavior of these parts, defined by the impact of quantum objects. He does not say, however, and does not mean that the independent behavior of quantum objects is described by quantum mechanics.

Bohr's elaboration implies, on the one hand, the unavoidable distinction, in quantum physics, between quantum *objects* and observed *phenomena*, and on the other, what he came to see as the *wholeness* of quantum phenomena, as discussed here. While, as I said, the difference between objects and phenomena is also found in classical physics, there we can, at least in principle, isolate objects from their interactions with measuring instruments, and treat observed phenomena as objects. In the case of quantum phenomena, at least in Bohr's interpretation, it is not possible to do so. In this respect, where we make the discrimination in question "between those parts of the physical system considered which are to be treated as measuring instruments and those which constitute the objects under investigation" would indeed not affect our predictions concerning the outcomes of quantum experiments. As Bohr's final sentence explains, however, this distinction is only arbitrary or "a matter of convenience" up to a point in the case of quantum phenomena. This sentence conveys the deeper physical meaning of Bohr's correspondence principle, as he understands the principle after quantum mechanics was introduced, and couples it to the mathematical meaning of the principle, defined by the fact that in this region the equations of classical mechanics would give the same predictions as those of quantum mechanics. What makes this point especially important is that quantum objects (including those parts of measuring instruments through which they interact with quantum objects) are always on the other side of the "cut" and may even be rigorously defined accordingly. It follows that quantum objects and processes can never be isolated from the measurement processes that give rise to quantum phenomena, eventually, again, understood by Bohr in term of his concept of phenomenon, which places quantum objects and processes beyond any description or even conception we can form. At the same time, our predictions, which only concern observable phenomena, are not affected insofar as the cut is made in the region where our quantum-mechanical and classical predictions coincide by the correspondence principle, and it cannot be made otherwise. The ultimate nature of the processes responsible for these phenomena is quantum, however, and these processes can never be extracted from phenomena, always indivisible or closed in this sense (Bohr 1954, PWNB 2, p. 73).

The features of Bohr's concept of phenomena described in the preceding discussion also lead Bohr to his associated concept of "atomicity," in the original

Greek sense of an entity that is not divisible any further, which, however, now applies at the level of phenomena, rather than referring, on Democritean lines, to indivisible physical entities, atoms, found in nature itself. The reason for this is that Bohr's argument just outlined enables him to transfer to the level of observable configurations manifested in measuring instruments all the key features of quantum physics—discreteness, discontinuity, individuality, and atomicity (indivisibility)—previously associated with quantum objects themselves. Both concepts, phenomena and atomicity, emerged in Bohr's work at about the same time and are more or less equivalent.

"This novel feature of atomicity in the laws of nature," was, according to Bohr, "disclosed" by "Planck's discovery of the quantum of action." Bohr sees this discovery as "supplementing in such unexpected manner the old [Democritean] doctrine of the limited divisibility of matter," although Planck's atomicity, too, was originally understood on these classical lines (Bohr 1938, PWNB 4, p. 38). I cite Bohr's initial formulation in the Warsaw lecture, which appears to be the first occasion of the appearance of this concept in print, but virtually identical formulations are found throughout his subsequent writings (e.g., Bohr 1949, PWNB 2, p. 33 and Bohr 1959, PWNB 3, p. 2). Like Bohr's concept of phenomenon, the concept of "atomicity" is defined in terms of certain *individual* effects of quantum objects upon the classical world, as opposed to Democritean atoms of matter itself. Thus, "atomicity" now refers to physically complex and hence *physically* subdivisible entities, and not to single physical entities, whether quantum objects themselves or point-like traces of physical events. These "atoms" are individual phenomena in Bohr's sense, rather than as indivisible atomic quantum objects, to which one can no longer ascribe atomic physical properties any more than any other kind of properties.[5]

Any attempt to "open" or "cut through" a phenomenon can only produce yet another closed individual phenomenon, a different "atom" or set of such "atoms," leaving quantum objects themselves irreducibly inaccessible inside phenomena. As Bohr says: "In fact, the individuality of the typical quantum effects finds its proper expression in the circumstance that any attempt at subdividing the phenomena will demand a change in the experimental arrangement introducing new possibilities of interaction between objects and measuring instruments which in principle cannot be controlled" (Bohr 1949, PWNB 2, pp. 40). "Consequently," he adds, thus also reinstantiating complementarity on these lines, "evidence obtained under different experimental conditions cannot be comprehended within a

---

[5] Among a few works that examine Bohr's concept of atomicity are Folse (1985, 1987) and Stapp (2007). Folse's interpretation of this concept or of Bohr's concept of phenomenon, and of Bohr's epistemology, is different from the one offered here. In particular, he attributes to Bohr a view that quantum mechanics is nonlocal, which is, as I argue, difficult to sustain. In fairness, Folse's (2002, p. 93) more recent commentaries no longer invoke nonlocality. The genealogy of the concept is not easy to trace, although one can find a few parallels, in particular, as noted by Stapp (2007), with Whitehead's (1929) concept of atomicity ("drops of experience"). It is unclear, however, whether Whitehead's ideas were familiar to Bohr or had impact on his thought.

single picture, but must be regarded as *complementary* in the sense that only the totality of the phenomena exhaust the possible information about the objects" (Bohr 1949, PWNB 2, p. 40). As a result, again, only probabilistic estimates concerning the outcome of our experiments can be given, since it is, in principle, impossible to reach the level of objects themselves, where one could encounter anything causal. That is why the situation simultaneously entails "the impossibility of subdividing quantum phenomena and reveal[s] the ambiguity of ascribing customary physical attributes to atomic objects" (Bohr 1949, PWNB 2, p. 51).

Each phenomenon is *individual*, each—every (knowable) effect conjoined with every (unknowable and even unthinkable) process of its emergence—unique and unrepeatable. Some of them can be clustered insofar as they refer to the "same" quantum entities, whether "individual" (for example, elementary particles) or collective (for example, more or less stabilized composites of quarks and gluons, such as protons or neutrons). Reciprocally, however, this view allows one to define and identify such entities, individually or collectively. Thus, along with quantum atomicity as *indivisibility,* quantum atomicity as *individuality* is now also understood as the individuality, and ultimately the uniqueness, of each phenomenon. By the same token, each phenomenon is discrete in the sense of being discontinuous in relation to every other such phenomenon, as are, for example, those associated with each trace (dot) on a silver screen left by a collision between a quantum object and the screen. Finally, there is yet another form of quantum *discontinuity*, that defined as the irreducible inaccessibility of quantum objects themselves, the impossibility of applying either of these concepts (continuity or discontinuity) or any conceivable concept to their "relation" (another inapplicable concept) to the manifested effects of their interaction with measuring instruments, which is responsible for these effects.

Bohr's concepts of phenomena and atomicity are new physical and philosophical concepts, and their conceptual architecture involves yet another novel epistemological concept, that of the irreducibly inconceivable nature of quantum objects and their behavior. At one level, the concepts of phenomena and atomicity do not involve any mathematics. They can, however, be related to the experimental data and mathematical formalism of quantum mechanics, insofar as the latter can predict these data, configured in terms of these concepts, which gives them a mathematical component. As full-fledge quantum-theoretical concepts, the concepts of phenomena and atomicity contain physical, mathematical, and philosophical components. By the same token, as I shall now discuss, the scheme is linked to probability in a way radically different from that of the old quantum theory, where the use of probability was linked to the Democritean view of quantum objects, as indivisible units of matter itself.

9.3 Bohr's Epistemology and Quantum Probability

Bohr's ultimate epistemology of quantum phenomena and quantum mechanics entails an understanding of randomness or chance (or correlations, such as those of the EPR type) and probability that is different from that found in classical physics, including in classical statistical physics. The difficulties of thinking concerning quantum probability in classical terms had begun to emerge already with Planck's black body radiation law, with Einstein the first to understand the seriousness of these difficulties. It may be argued, however, for the reasons discussed in Chap. 3 that the history of the type of understanding found in Bohr's interpretation, especially in its ultimate version, begins with Heisenberg's introduction of quantum mechanics as a theory dealing with the probabilities of transitions between stationary states of electrons in atoms. Shortly thereafter, following Schrödinger's introduction of his wave mechanics, Born's probabilistic interpretation of the wave function and Born's rule for deriving probabilities from amplitudes extended Heisenberg's approach to quantum mechanics as a whole.

Bohr's understanding of quantum randomness (or correlations) and quantum probability is defined by the suspension of the applicability of the idea of causality even to individual quantum processes and events. As explained earlier, the concept of event is especially pertinent here, since this individuality, too, is only manifested at the level of observable events, or again, phenomena, and not at the level of quantum objects and processes, to which the concept of individuality may not rigorously apply any more than any other concept. In this sense, the elementary particle of a given type, such as the electron, is defined by a given set (potentially very large, but quite specific) of possible phenomena or events observable in measuring instruments associated with it. This set is the same for all possible electrons, while the correlation between any such phenomenon and any given electron can never be assured, which makes us speak of electrons as indistinguishable from one another. There are also sets of phenomena associated with multiple particles of the same type, also with large multiplicities of particles as in quantum statistics. As we have seen, this situation becomes especially dramatic in high-energy regimes, where we need quantum field theory. This understanding is, thus, fundamentally different from the view adopted in classical statistical physics, where the behavior of the individual entities comprising the multiplicities considered statistically is assumed to be causal, at least ideally. In Bohr's ultimate view, our predictions themselves only concern the effects of such processes manifest in the measuring instruments involved. As we have seen, this was in effect the case already in Heisenberg's paper introducing quantum mechanics. This is why, as I argue here, Heisenberg's paper was important for Bohr not only initially but also throughout his work.

As stressed throughout this study, the probabilistic character of our predictions concerning quantum phenomena is unavoidable, since, to return to Bohr's formulation, "one and the same experimental arrangement may yield different recordings" (Bohr 1954, PWNB 4, p. 73). It is, again, possible to speak of "one and the same experimental arrangement," because, unlike the outcomes of experiments, we can control the measuring instruments involved, given that the parts of these instruments relevant for setting up our experiments can be described classically. Under these conditions, the probabilistic character of such predictions will also concern primitive individual quantum events. For, unlike in the case of certain classical individual events, such as a coin toss, in the case of quantum phenomena it does not appear possible—and in Bohr's ultimate view, it is in principle impossible—to subdivide these phenomena into entities of different kinds, concerning which our predictions could be exact, even ideally or in principle. Any attempt to do so will require the use of an experimental setup that leads to a phenomenon or set of phenomena of the epistemologically same type (they could be different physically), concerning which we could again only make probabilistic predictions.

Accordingly, rather than whether quantum phenomena entail a probabilistic theory predicting them (since they manifestly do), the question is whether there is or not an underlying classical-like causal dynamics ultimately responsible for such events. If this kind of underlying dynamics exists, it would imply that a classical-like account of such events could in principle eventually be developed, as Einstein hoped with the model of classical statistical physics in mind, although it cannot be guaranteed that it would. By contrast, Bohr argued that quantum phenomena appear to disable, and in his ultimate interpretation they do, the underlying assumptions of classical statistical physics (or accounts of individual classical phenomena that we cannot sufficiently track), based on the causality of the primitive (indecomposable) individual processes involved. In other words, in quantum mechanics and higher level quantum theories, it is difficult to assume that randomness or correlations and, hence, the necessity of using probability merely arise in view of our inability to access the underlying *causal* dynamics determining the behavior of the system considered. Instead, in Bohr's view the irreducible role of randomness and probability in quantum physics arises as consequences of "the inability of the classical frame of concepts to comprise the peculiar feature of indivisibility, or 'individuality,' characterizing the elementary processes" (Bohr 1949, PWNB 2, p. 34). These features are, again, captured by Bohr's concepts of phenomenon and atomicity, which entail the irreducibly unthinkable nature of quantum objects and processes. As noted from the outset of this study, the lack of causality is an automatic consequence of this epistemology, since causality would imply that we could think this nature, however partially. The fact that probabilistic predictions of quantum mechanics and higher level quantum theories are correct is, again, enigmatic, insofar as it does not appear to have an underlying physical justification of the type found in classical statistical physics (in Bohr's ultimate view, it does not), and we appear to be lucky to be able to make them.

As explained in Chaps. 4 and 5, Bohr begun to move toward "the renunciation of the ideal of causality" under the impact his exchanges with Einstein, following the Como lecture, which still retained, if uneasily, this ideal. Bohr appears to have arrived at his ultimate view on the subject, along with (and, again, as a consequence) of his ultimate epistemology, by the time of "Complementarity and

Causality" and the Warsaw lecture. Bohr's exchanges with Einstein, now concerning the EPR experiment, again played a key role in this development of Bohr's thinking. "Complementarity and Causality" clearly establishes the connections between Bohr's ultimate epistemology and "the renunciation of the ideal of causality:" "The renunciation of the ideal of causality in atomic physics which has been forced on us is founded logically only on our not *being* any longer in a position to speak of the autonomous behavior of a physical object" (Bohr 1937, PWNB 4, p. 87). "Complementarity and Causality" does not yet see this situation in terms of the concepts of phenomenon and atomicity. But the Warsaw lecture, "The Causality Problem in Atomic Physics," does, indeed in its opening paragraph:

> The unrestricted applicability of the causal mode of description to physical phenomena has hardly been seriously questioned until Planck's discovery of the quantum of action, which disclosed *a novel feature of atomicity* in the laws of nature supplementing in such unsuspected manner the old doctrine of the limited divisibility of matter. Before this discovery statistical methods were of course extensively used in atomic theory but merely as a practical means of dealing with the complicated mechanical problems met with in the attempt at tracing the ordinary properties of matter back to the behavior of assemblies of immense numbers of atoms. It is true that the very formulation of the laws of thermodynamics involves an essential renunciation of the complete mechanical description of such assemblies and thereby exhibits a certain formal resemblance with typical problems of quantum theory. So far there was, however, no question of any limitation in the possibility of carrying out in principle such a complete description; on the contrary, the ordinary ideas of mechanics and electrodynamics were found to have a large field of application also proper to atomic phenomena, and above all to offer an entirely sufficient basis for the experiments leading to the isolation of the electron and the measurement of its charge and mass. Due to the essentially statistical character of the thermodynamical problems which led to the discovery of the quantum of action, it was also not to begin with realized, that the insufficiency of the laws of classical mechanics and electrodynamics in dealing with atomic problems, disclosed by this discovery, implies a shortcoming of the causality ideal itself. (Bohr 1938, PWNB 4, pp. 94–95; emphasis added)

Bohr, thus, makes a strong historical claim, which he elsewhere extends to the history of philosophy as well. This extension is not surprising, given the fundamental relationships between classical physics (first, classical mechanics and then classical statistical physics) and the philosophy of causality, which preceded classical physics and shaped it conceptually, but was then, from Kant on, even more significantly shaped by classical physics. According to Bohr:

> [E]ven in the great epoch of critical [post-Kantian] philosophy in the former century, there was only a question to what extent a priori arguments could be given for the adequacy of space–time coordination and causal connection of experience, but never a question of rational generalizations or inherent limitations of such categories of human thinking. (Bohr 1949, PWNB 2, p. 65)

As noted earlier, even the more radical philosophical questionings of causality, such as those by Kant or still more skeptical Hume, are those of our incapacity to perceive the underlying causal world, which would be presupposed at the ultimate level as inaccessible to us.

In physics (one is compelled to agree with Bohr), this more radical questioning of causality does not appear before quantum physics comes on stage to reveal that the lack of causality may be due to "the inability of the classical frame of concepts to comprise the peculiar feature of indivisibility, or 'individuality,' characterizing the elementary processes," the unthinkable quantum processes that result in this indivisibility and individuality (Bohr 1949, PWNB 2, p. 34). One should perhaps refer to the indivisibility *and* individuality of phenomena, restricting us to probabilistic estimates even as concerns the outcome of individual quantum events, something that Einstein refused to entertain. The main point, however, is that the lack of causality rather than only determinism in the case of quantum phenomena (in the usual sense) is an automatic consequence of Bohr's interpretation of these phenomena in terms of his concepts of phenomena and atomicity. Unlike merely the unknowable but in principle thinkable character of Kant's noumena or things-in-themselves, which still allows causality at the ultimate level, the irreducible unthinkable nature of quantum objects and processes leaves no room for causality. For causality would, again, be a way of conceiving, even if partially, of these processes, would make them at least partially thinkable, and, to return to Wittgenstein's observation, cited in his introduction, conversely, we may not be able to think of a process otherwise than causally (Wittgenstein 1924, p. 175).

There is still the question of quantum correlations, the "enigma" of correlations, defined by the circumstance that, while each individual quantum event is *always* irreducibly lawless, in certain circumstances multiple quantum events exhibit statistical correlational orders, such as that found in the EPR-type experiments. The presence of these correlations is, however, consistent with the unthinkable and hence noncausal character of quantum processes, and it appears to have served as a further impetus for Bohr's ultimate interpretation of quantum phenomena and quantum mechanics. I am, again, not saying that correlations would necessitate this type of interpretation, although Bohr (without expressly speaking of correlations) comes close to making this type of argument. In any event, the enigmatic combination of individual lawlessness of each individual quantum event and statistically correlated nature of certain aggregates of such events leaves space for probabilistic predictions, space that quantum mechanics appears to use maximally. But, in Bohr's ultimate view, it leaves little, if any, space for the description or even conception of the actual processes responsible for this situation, and, again, as a consequence for the underlying causality behind this combination.

9.4 Einstein and Bohr on Locality and Probability in Quantum Mechanics

Bohr's and Einstein's epistemological positions may be seen as correlative to their respective views of probability in quantum physics. The question of locality of quantum mechanics is also implicated in the statistical aspects of the EPR situation, addressed in Einstein's later arguments of the EPR type, which I would like to consider now. First, it might be helpful to briefly reprise the key relevant points of my discussion in the preceding chapter. EPR's argument is, Bohr contends, made insufficient by the nature of quantum phenomena as defined by the irreducible role of measuring instruments in their constitution, which makes it impossible to consider the behavior of quantum objects independently of their interaction with these instruments. The application of EPR's criterion of reality becomes ambiguous by virtue of the lack of the qualifications of this criterion required by these conditions, not adequately considered by EPR. "Since," Bohr concludes, "these conditions constitute an inherent element of the description of any phenomenon to which the term 'physical reality' can be properly attached, we see that the argument of the mentioned authors does not justify their conclusion that quantum-mechanical description is essentially incomplete" (Bohr 1935, p. 700). These conditions and, hence, the completeness of quantum mechanics are also essentially linked to the irreducibly probabilistic nature of our predictions concerning quantum events, in particular, individual quantum events. This circumstance may not appear relevant in the EPR situation, since the predictions in question can be made with probability equal to unity. As I argued in Chap. 8, however, it is crucial there, and is merely obscured by the fact that the EPR (idealized) predictions are made with certainty. The role of these statistical considerations appears to have been underappreciated by EPR, although it did not escape Einstein's attention in his later arguments. EPRs and these later arguments by Einstein for the incompleteness, or else nonlocality, of quantum mechanics subtly depend on the proper coordination of the outcomes of the key measurements involved in two repeated EPR experiments on two different, but identically prepared, EPR pairs. This type of coordination is possible in considering classical phenomena or (this difference is not relevant in classical physics) objects. It is, however, not possible in considering quantum phenomena, including those of the EPR type, in view of the impossibility of securing the identical outcomes of the identically prepared experiments. It follows that quantum mechanics, which adequately accounts for what is possible to account as concerns these phenomena, may be seen as a complete, as well as (since an analogous counterargument to EPR applies in the case of locality) local, theory of quantum phenomena. At least, EPR have not demonstrated otherwise. Bohr, accordingly, contends that, rather than entailing the incompleteness or nonlocality of quantum mechanics, the situation entails "the necessity of a final renunciation of the classical ideal of causality and a radical revision of our attitude toward the problem of physical reality," a revision that allows the locality of quantum phenomena and quantum mechanics (Bohr 1935, p. 697).

It is this joint imperative that led Bohr to his ultimate view, expressed in the statement that just about defines his response to Einstein's criticism of quantum mechanics. He says: "in quantum mechanics, we are not dealing with an arbitrary renunciation of a more detailed analysis of atomic phenomena, but with a recognition that such an analysis is *in principle* excluded" (Bohr 1949, PWNB 2, p. 62). Bohr specifically responds here to Einstein's argument made in 1936 in

"Physics and Reality," described by Bohr as follows: "[T]he quantum-mechanical description is to be considered merely as a means of accounting for the average behavior of a large number of atomic systems," as against "the belief that it should offer an exhaustive description of individual phenomena" (Bohr 1949, PWNB 4, p. 61). This argument persists throughout Einstein's life; apart from using it in the Schillp volume (Einstein 1949a, b), Einstein invokes it, for example, in his letters to Born as late as 1954 (Born 2005, pp. 166–170, 204–205, 210–211). In Einstein's words, he saw "the belief that it should offer an exhaustive description of individual phenomena" as "logically possible without contradiction," but found it "so very contrary to [his] scientific instinct that [he could not] forego the search for a more complete conception" (Einstein 1936, p. 375; Bohr 1949, PWNB 4, p. 61).

It is not altogether clear how far Einstein perceived the implications of Bohr's ultimate view of the situation, which Bohr adopts in commenting on Einstein's argumentation. Einstein might have seen Bohr's or related interpretations along the lines of Bohr's reply to EPR, whereby, in accordance with the uncertainty relations, one of the two conjugate quantities is assumed ascribable to a quantum object at the time of a measurement. On the other hand, as discussed earlier, he also misread Bohr by assuming that his position allows nonlocality. In any event, Bohr's ultimate view was not quite yet in place in 1936 when Einstein's article was published. By contrast, Bohr's 1949 comments, in response to Einstein's statement, implies and is based on his ultimate view of quantum mechanics, established after Einstein's article. Bohr, thus, appears to suggest that Einstein rejects his ultimate interpretation as well, which would not be surprising. In any event, Bohr's statement in question applies either way, given that even in his interpretation in his reply to EPR there could be no question of a further analysis, of the type desired by Einstein, an analysis that would allow a description, especially a causal description, of individual quantum processes considered independently of measurement.

The question is on what grounds a search for an alternative may be deemed successful, especially if the EPR-type experiment may be shown not to offer such grounds. For a mere *rejection* of given argumentation (such as that of Bohr) and a desideratum for an alternative conception (such as the one Einstein has in mind), do not constitute a demonstration of either logical or experimental deficiency of this argumentation, even though, as Bohr notes, Einstein's "attitude might seem well balanced in itself" (Bohr 1949, PWNB 4, p. 62). According to Bohr: "In my opinion, there could be no other way to deem a logically consistent mathematical formalism as inadequate than by demonstrating the departure of its consequences from experience or by proving that its predictions did not exhaust the possibilities of observation, and Einstein's argumentation could be directed to neither of these ends" (Bohr 1949, PWNB 2, p. 57). Ostensibly, this statement, again, refers to Einstein's earlier arguments, but, by this time, in 1949, Bohr clearly had all of Einstein's arguments in mind, including those of the EPR type. Indeed, Bohr follows this sentence with his point that grounded my discussion of the EPR experiment in the preceding chapter, that "we must realize that in the problem in question we are not dealing with a *single* specified experimental arrangement, but are referring to *two* different, mutually exclusive arrangements" (Bohr 1949, PWNB 4, p. 57; Bohr's emphasis).

There are further subtleties to Einstein's 1936 assessment that the belief in question is "logically possible without contradiction... but [is] so very contrary to [his] scientific instinct." It is true that this belief refers, as Bohr says, to the argument that "the quantum-mechanical description is to be considered merely as a means of accounting for the average behavior of a large number of atomic systems." But it also refers to the belief that one cannot derive the inner workings of individual quantum systems from quantum mechanics as a probabilistic theory, or, in Einstein's view (not shared by Bohr) a statistical theory of quantum ensembles. It replies to the following rhetorical question: "Is there really any physicist who believes that we ever [from the quantum-mechanical way of dealing with the situation] get an inside view of these important alterations [due to individual perturbations] in the single system, in their structure and their causal connections, and this regardless of the fact that these single happenings have been brought so close to us, thanks to the invention of the Wilson [cloud] chamber and the Geiger counters?" (Einstein 1936, p. 375). Einstein's question is not unreasonable. It is not easy to believe that it is possible to ever get an inside view, also literally in the sense of visualization [*Anschaulichkeit*], of the inner workings of individual quantum systems themselves from quantum mechanics as only a probabilistically predictive theory. Einstein's invocation of "causal connections" is telling here. Indeed, it is, by definition, impossible to get such "an inside view" in Bohr's interpretation, even in the version found in his reply to EPR and more radically in his ultimate version of it. One is likely to need a different theory in order to achieve this, if—the big if, Bohr argued—quantum phenomena will ever allow us to have such a theory. Einstein believed that such a theory should be possible in view of the EPR experiment, while Bohr argued—including by way of exposing the problems of EPR's argument and related arguments by Einstein— that it might not be possible. This view appears to be supported by Bell's theorem and related developments, or at least, thus far, it is not contradicted by either theoretical arguments or experiments. As I noted earlier, definitive claims are difficult in this domain, especially when it comes to Bell's theorem.

Some of these arguments do allow the possibility that quantum mechanics could be regarded as a statistical theory of ensembles, in which case, he also argued, it could be seen as local. In this view, quantum mechanics would be *analogous* to classical statistical physics. According to Einstein, if one regards the wave function as relating to "many systems, to 'an ensemble of systems,' in the sense of statistical mechanics," then "the paradox" arising in view of EPR's argument is eliminated and quantum mechanics could be seen as local (Einstein 1936, p. 375; Einstein 1948; Einstein 1949a, p. 81; Letter to Born, 3 Dec 1953, Born 2005, p. 205; Letter to Born, 12 January 1954, Born 2005, p. 211). Einstein's statistical alternative, however, still leaves quantum mechanics incomplete by his criterion by virtue of its inability to provide a properly exhaustive physical description of the behavior of individual quantum systems of the kind classical mechanics does, including for the individual constituents (such as molecules of a gas) of the systems considered in classical statistical physics. At most, quantum mechanics provides an incomplete description of such systems along the lines of EPR's argument, since one must maintain locality at this level. "Physics and Reality" makes an even stronger statement: "what happens to a single system" is also "*entirely* eliminated" (Einstein 1936, pp. 375–377; emphasis added). In other words, in this view, quantum mechanics provides no account of individual quantum systems at all. On subsequent occasions, Einstein, again, appears to assume that quantum mechanics gives an *incomplete* description of individual systems on the lines of EPR's article, which, again, allows one to avoid nonlocality. As he says, "one can safely accept the fact … that the description of the single system is incomplete, if one assumes that there is no corresponding complete law for the complete description of the single system which [law] determines its development in time" (Letter to Born, 3 December 1953, Born 2005, p. 205). In this case: "The statistical character of the … theory would … follow necessarily from the incompleteness of the description of the [individual] systems in quantum mechanics" (Einstein 1949a, p. 81). Thus, from Einstein's point of view, a proper mechanics is lacking either way. If quantum mechanics is a complete description of individual quantum systems, it is nonlocal, which is unacceptable; it may be considered local if it is merely a statistical theory of ensembles, in which case, however, there is either an incomplete description of individual quantum systems or no such description at all.

Bohr counters that one need not conclude, as Einstein does, that quantum mechanics, if, again, local, could be seen only as "*an essentially statistical theory of ensembles.*" It may instead be seen as *a fundamentally probabilistic theory of individual quantum systems* or *individual quantum phenomena.* Large quantum ensembles are covered by a separate quantum theory, quantum statistics, to the development of which Einstein made an important contribution, via Bose's theory (Einstein 1925). Einstein is right insofar as quantum mechanics does not provide a description of the independent individual behavior of quantum systems, considered apart from their interaction with the measuring instruments. This is true even in the less radical version of Bohr's argument offered in his reply to EPR. In his ultimate interpretation, quantum systems and their behavior are, again, left altogether beyond all description or conception even at the time of measurement. Either way, Bohr's interpretation makes quantum mechanics a probabilistic theory of individual quantum process and phenomena to which they give rise.

A causal or realist theory of these processes cannot, again, be excluded, and Bohr was cautious in this respect, as suggested by his persistent use of such locutions as "would seem" and "would appear." As I have stressed throughout this study, it would be difficult to make a long-term guess. For now, quantum mechanics (or by implication higher level quantum theories) in the interpretation proposed by Bohr is, at least, good enough, and, in the view of the present author, at the least as good as any alternative (known to this author) and better than most. "Skepticism about the necessity of going that far in renouncing the customary demand as regards the explanation of natural phenomena," invoked by Bohr in 1949, continues to be widespread (Bohr 1959, PWNB 2, p. 63). One can, however, also take a more positive view of the situation. Quantum mechanics and higher level quantum theories allow us to make excellent predictions concerning experimental observable quantum phenomena. This fact enables the disciplinary practice of quantum mechanics and higher level quantum theories as branches of modern physics as a mathematical–experimental science of nature that it has been since Galileo. That they do so without offering a description of quantum objects or their behavior should not perhaps be held against them, given that it is possible that nature precludes us from having such a description and even forming any conception concerning quantum objects and processes. If such is the case, the capacity of quantum theories to probabilistically predict the outcomes of quantum experiments makes them all the more remarkable and makes us fortunate to be able to invent them.

9.5 Postscript on Complementarity and Bohr's Epistemology Beyond Physics

My main focus in this book has been on Bohr's thinking concerning physics, even when this thinking deals with philosophical questions, a focus defined by the main aim and the scope of the book's project. I would also argue, however, that Bohr was at his best philosophically when dealing with physics, rather than in extending his concepts, such as complementarity, and arguments beyond physics, which, as he always qualified, he was doing in tentative and preliminary ways. Nevertheless, throughout his writings, Bohr invokes possible extensions of complementarity (which is defined very generally, thus allowing for such extensions) and his epistemology to other fields. He clearly believed these connections to be viable, and volume 10 of his collected works, appropriately entitled *Complementarity Beyond Physics*, devoted to the subject, contains a large amount of published works and (previously) unpublished archival material (Bohr 1972–1999, v. 10). Accordingly, I would like to briefly address some of such connections here, focusing primarily on Bohr's epistemology, especially in its ultimate form, rather than on the concept of complementarity, more commonly considered commentaries on Bohr in this context. I would argue that, although expressly suggested by Bohr, the extensions of the concept to biology and psychology are more tentative and uncertain in their effectiveness, especially in the case of biology, in part in view of more recent developments there, many of which came during the last four decades, after Bohr's death. On the other hand, Bohr's general epistemology, especially in conjunction with probability, appears to me to have a great potential for effective applicability in several fields, biology and psychology among them. I would like, however, to consider one example of Bohr's attempt to extend the concept of complementarity to psychology, which is connected to the type of epistemology that defines his ultimate view of quantum phenomena and quantum mechanics, as considered in this chapter. Bohr writes:

An especially striking example [of complementary experiences in psychology] is offered by the relationship between situations in which we ponder on the motives for our actions and in which we experience a feeling of volition. In normal life, such shifting of the separation is more or less intuitively recognized, but symptoms characterized as "confusion of the egos," which may lead to dissolution of the personality, are well known in psychiatry. The use of apparently contrasting attributes referring to equally important aspects of the human mind presents indeed a remarkable analogy to the situation in atomic physics, where complementary phenomena for their definition demand different elementary concepts. Above all, the circumstance that the very word "conscious" refers to experience capable of being retained in the memory suggests a comparison between conscious experiences and physical observations. In such an analogy, the impossibility of providing an unambiguous content to the idea of subconsciousness corresponds to the impossibility of pictorial interpretation of the quantum-mechanical formalism. Incidentally, psychoanalytic treatment of neuroses may be said to restore balance in the content of the memory of the patient by bringing him new conscious experiences, rather than by helping him to fathom the abysses of his subconsciousness. (Bohr 1954, PWNB 2, p. 77)

It is difficult to ascertain to what degree Bohr's particular argument here is workable in psychology, and specifically in psychiatry, or psychoanalysis, and it is not that important here.[6] My point instead is the possibility of applying Bohr's concept of complementarity and his epistemology (both are clearly linked here, which is my point as well), and such applications may, I would argue, prove to be effective in psychology or elsewhere, including in biology and philosophy, two fields that are most often invoked by Bohr in this connection.

This possibility should not be surprising, given the history of certain epistemological problems considered in these fields, which exhibit marked affinities with epistemological problems of quantum theory, especially from the eighteenth century Enlightenment on, although, as Bohr notes, one could trace this history much earlier, in some respects, even to the pre-Socratics. First of all, the Enlightenment and its aftermath coincide with several major developments that posed new questions concerning reality and causality for science, philosophy, and the literature, three such developments in particular. The first is the advent of a new type of philosophical thinking, reflecting a more skeptical attitude toward

---

[6] K. G. Jung and others tried to use the concept of complementarity in psychology and psychoanalytic theory. Jung extensively corresponded with Pauli (who was in a Jungean analysis for a while). Complementarity was discussed in this correspondence, although along the lines of Pauli's understanding of the concept and his epistemological views, rather than those of Bohr. See Pauli's account, which also addresses more general epistemological connections between quantum theory and psychoanalytic theory (Pauli 1994, pp. 149–164). See also (Gieser 2005) and articles on these connections in (Atmanspacher and Primas 2009). Some commentators also pursue connections between Bohr's concept of complementarity and William James's concept of "*complementary* consciousnesses," on the account of the role of mutual exclusivity in both concepts (James 2007, p. 206). Although Bohr briefly mentions James in positive terms, the degree of Bohr's familiarity with James's work remains unclear, and there is no strong evidence that he was aware of James's use of "complementary" (as an adjective) in this context. More importantly, the overall architecture of Bohr's concept and his epistemology are , in my view, different from those of James. One might argue for closer connections between Bohr's epistemology and Nietzsche's, Freud's, and more recently, Jacques Lacan's psychological and psychoanalytic epistemology. I can, however, only mention these connections here without addressing them.

causality, in Hume's and then Kant's philosophy, and as noted in the Introduction, Kant was primarily responsible for the term "Enlightenment." The second is the emergence of probability theory, which had major implications for mathematics, science, philosophy, and the culture of modernity. The third has to do with eighteenth century optics, as the site of the major debate concerning the corpuscular versus wave nature of light. It may be added that the philosophy of Hume and Kant, their followers, and the Romantic literature of the period had influenced the thought of such scientific figures as Darwin, Maxwell, and Boltzmann, and later on Heisenberg and Bohr. Maxwell and Boltzmann were responsible for key developments that eventually led to quantum-theoretical thinking, via Planck's law, which was formulated at the intersection of thermodynamics, electrodynamics, and, for the first time, quantum theory, and which was statistical. Quantum theory revived the debates concerning the wave versus particle nature of light—admittedly, in a hitherto unprecedented form, but not without philosophical affinities with earlier stages of these debates.

If it may be unexpected that this type of *thinking* (some of the *questions* to which this thinking responds are, again, much older) have appeared in philosophy, literature, and art even before quantum physics, that this thinking had entered the modernist literature and art, or in philosophy, in the wake of quantum mechanics cannot come as a surprise. A number of key modernist literary figures could be invoked here, Franz Kafka, for example. His work could be expressly connected to the old quantum theory (that of Planck, Bohr, A. Sommerfeld, and Einstein), although not to quantum mechanics itself, which was introduced after Kafka's death (in 1918). Kafka's works, especially his novels, *The Trial* (*Das Prozess*), *America*, and *The Castle*, refer to phenomena or processes and places, such as those named by these titles, whose ultimate nature or, one might say, structure— "topology" and "geometry"—are ultimately unknowable and even inconceivable, quite similarly to Bohr's view of quantum objects and processes.[7] Joyce's *Ulysses* and *Finnegan's Wake*, Robert Musil's *The Man without Qualities*, and Samuel Beckett's plays also offer instances of quantum-like thinking, sometimes, coming close to Bohr's ultimate epistemology (Plotnitsky 2012). Similar affinities with quantum-theoretical thinking are found in modernist music and visual arts, clearly influenced by quantum theory from Planck on. Bohr's thinking has been associated with cubism by several commentators. As Pais reports, "among the paintings hanging in [Bohr's] house was one by [one of the founders] of cubism [Jean Metzinger]" (Pais 1991, p. 335). By contrast, Einstein, again, according to Pais did not appear to care for cubist strange imagery, any more than he cared for the strangeness of quantum mechanics. A classicist in everything, Einstein, an accomplished violinist, preferred Haydn to the like of Schoenberg and Stravinsky and their quantum-like harmonies, which would even transform, as in some modernist works, the classical continuity of the violin into the quantum percussiveness of the piano. Bohr played piano (not very well, it appears, in

---

[7] See, for example, (Derrida 1992).

contrast to Heisenberg, an accomplished piano player), and he must have thought of this, to be sure, as a coincidental, but still intriguing contrast between his and Einstein's musical instruments.

For the moment, according to Bohr, and this point is crucial in the present context, the main reasons for the affinities between quantum-theoretical thinking and thinking in other fields is the shared nature of certain problems that one confronts in different fields and, as a result, of *some* responses to these problems. One can of course also respond differently to the same problems, and might want a different response, as again, Einstein did in the case of the quantum puzzle, as he sometimes called it. Bohr says: "We are not dealing here with more or less vague analogies, but with an investigation of the conditions for the proper use of our conceptual means of expression. Such considerations not only aim at making us familiar with the novel situation in physical science, but might ... be helpful in clarifying the conditions for objective description in wider fields" (PWNB 2, p. 2). This reprises an earlier statement, which nuances his point: "we are not dealing with more or less vague analogies, but with clear examples of logical relations which, in different contexts, are met with in wider field" (Bohr 1958, PWNB 3, p. 7).

Questions concerning these conditions and these relations have, however, sometimes been posed more radically and closer to the way quantum theory posed them, in philosophy or the literature and art than in science prior to quantum theory, especially in the literature and art. This, I would argue, was because the literature and art confronted certain troubling questions concerning human life more directly than philosophy did, questions often linked to the difficulty or even impossibility of relying on the ideas of reality and causality. Although there are (rare) exceptions, in general philosophers have been more reluctant to doubt that some underlying causal ontology ultimately governs nature and mind, although most philosophers have recognized the practical difficulties of capturing the actual character of this ontology. In this respect, the situation is not so different, and in turn, connected to the response to quantum mechanics and Bohr's epistemology, especially in its ultimate form, considered in this chapter, on the part of most physicists, beginning, again, with Einstein. It is not that the same type of resistance is not found in the literature and art, but it appears that certain works there managed to operate against the classical, realist and causal, model of thinking and representation for quite a while with a greater degree of freedom and permissiveness than elsewhere.

It is true that, even leaving aside mathematics, there are limits to philosophical parallels, such as those invoked by Bohr, between quantum mechanics and other fields, and we must rigorously respect these limits. Bohr never believed otherwise. Quantum phenomena and quantum mechanics have their specific, even unique features, some of which would have been difficult and even impossible to imagine if nature itself did not show them to us. In Wheeler's words, "What could one have dreamed up out of pure imagination more magic—and more fitting—than this?" (Wheeler 1983, p. 189) Thus, it may also be true, at least in some respects, that, as the particle physicist S. Coleman (reportedly) said, "if thousands of philosophers spent thousands of years searching for the strangest possible thing, they would never find anything as weird as quantum mechanics" (reported in Randall 2005, p. 117). On the other hand, philosophers, beginning with the pre-Socratics, or poets, have spent thousands of years exploring things that are pretty weird, or in any event, quantum mechanical-like, as Bohr clearly realized, as must be apparent from the discussion just given. Indeed he expressly spoke of "the epistemological situation here encountered [i.e. in the case of quantum phenomena], which *at least in physics* is of an entirely novel character" (Bohr 1937, PWNB 4, p. 87; emphasis added). This suggests that he thought that, at least in some respects, this situation might have been previously encountered elsewhere. Human affairs, the main source of philosophical and poetic thinking (although physics has also served as such a source), are often as strange as in quantum physics. Have not Shakespeare, Dostoyevsky, Joyce, Kafka, and Beckett told us as much? One need not search for the strangest possible things, one cannot avoid them. It is true that the strangeness of quantum phenomena leaves room for rigorous, mathematically defined laws, without, it appears, leaving room for explaining what is behind these laws. This may be strange. But is not our capacity to think of these laws just as strange?

To illustrate this point, I would like to sketch (following a previous analysis in Plotnitsky 2012) a possible interpretation of Beckett's *Endgame* as providing a depiction of quantum-like probabilistic reasoning, although in this case the influence of quantum physics on Beckett's thinking is palpable, and this kind of depiction would indeed be unusual before quantum mechanics. Given, however, that my main aim here is to illustrate the possibility of the epistemology of the type developed by Bohr beyond quantum mechanics, this is no disadvantage. The example of Beckett's plays may also be fitting here given Bohr's famous appeal to a theatrical metaphor that he used on several occasions: "the new situation in physics reminded us of the old truth that we are both onlookers and actors in the great drama of existence" (PWNB 1, p. 119; also PWNB 2, pp. 20, 63). Bohr does not elaborate on the specific nature of this resemblance, thus, leaving the statement open to an interpretation, and a few have been offered by commentators, specifically in relation to psychological processes, such as introspection or self-reflection, connections that I shall not pursue here. On the other hand, Bohr might also have had in mind the following important feature of quantum experiments mentioned earlier in this chapter. In quantum experiments, our decision concerning what we want to do will only allow certain outcomes (for example, those concerning a future position measurement) and will unavoidably exclude the possibility of certain other, complementary types of outcomes (those concerning a future momentum measurement). In this sense, we are actors. On the other hand, we are spectators insofar as this freedom only allows us to select and control the setup of the experiment but not its outcome. This situation finds its parallels beyond physics, and it is especially pertinent to what takes place in Beckett's plays, as part of a broader epistemology that they share with quantum mechanics and Bohr's interpretation, especially in its ultimate version.

Let us recall, first, that in Bohr's view, quantum mechanics is a *probabilistic* theory of *individual* quantum processes or events. We can predict the probabilities of such events and only then by using the equations of quantum mechanics and particular rules of using those, such as Born's rule, based on his reinterpretation of Schrödinger's wave function in terms of probabilities. In this interpretation, the concept of wave is no longer given a physical meaning but only refers, meta-phorically, to the way probabilities of our predictions would "propagate" depending on the point to which a prediction would refer (Born 1926). Consider an individual quantum object, say, an electron, whose initial position is specified by a measurement. In Bohr's ultimate view, the corresponding Schrödinger equation allows one to predict the probability of finding the electron in a given region of space at a future time, say, in one second, without *describing* the behavior of the electron itself between these two experiments, one already performed and one to be performed. In other words, unlike in classical physics, once we (as "actors") make a measurement of the position of an electron at a given point, which we can do exactly, we cannot say where the electron will exactly be at a later point. We can only estimate a probability that it will be in a certain region of space, and there is always a nonzero probability that it will not be found anywhere. As concerns what actually happens, we can only be spectators on the stage of the experiment. After one makes a measurement, one might speak of certain *potential* or *virtual* probabilities for each given subsequent point in time at which a new measurement *could be* in principle performed. We can compile what Schrödinger called a cat-alogue of probabilities for predicting the outcome of experiments possibly to be performed at each such point (Schrödinger 1935, p. 158). It is in this sense that Born referred (metaphorically) to the wave-like propagation of probabilities in quantum mechanics (Born 1926, p. 804).

Now, comes the most crucial point of my discussion at the moment, which we have encountered in one of Bohr's elaborations considered in Chap. 5: "a sub-sequent measurement to a certain degree deprives the information given by a previous measurement of its significance for predicting the future course of the phenomena. Obviously these facts not only set a limit to the *extent* of the infor-mation obtainable by measurements, but they also set a limit to the *meaning* which we may attribute to such information" (Bohr 1929b, PWNB 1, pp. 17–18). That is, any *act* of measurement (we are, again, actors each time) discontinuously resets both the future evolution of the system and the propagation of virtual probabilities following this measurement. In other words, the process starts anew with each new measurement, which erases the outcome of the previous measurements (which we know) as meaningful for future predictions concerning the system. Each mea-surement changes the expectation catalogue, or rather creates a new one. Indeed, after a predicted measurement is performed, it may not be possible to use the object for any subsequent predictions at all. A new initial experiment upon a new object, and a new sequence of potential future measurements, will have to be set up. Quantum mechanics is about the future and the uncertainty of the future.

Under these conditions, the "game" begins anew each time, and there is, as Beckett, I think, realized in *Endgame*, no endgame, no-*end* endgame, in the sense of a possible causally determined trajectory of meaning (Beckett 1981). In chess, to which Beckett's title refers as well, the endgame, *endspiel*, of each game or even in general could, in principle, be determined, since chess is a zero-sum game with complete information and hence has a determined outcome. The current consensus is that the white wins, although that has not been proven. In principle, if we had sufficiently powerful computers, two such computers would always play the same game. Conceptually or logically, the game of chess would be finished, although one could of course still play it. When humans play the game it is as much a matter of quantum-like resetting of probability spaces, since human factors diminish or destroy a formal inevitability of the outcome.

The same type of belief in the causal enclosure of an event is perhaps also the fallacy of those "waiting for Godot" in the play under this name (Beckett 2008). The word "waiting" often implies a certain causality of an event coming from somewhere according to determinate expectation, perhaps ultimately determined by God, or by the second coming of God. In contrast, in Beckett it is a coin toss, a throw of dice. Accordingly, reversing the proverbial expression, it is always the question of *if*, and never of *when*. Only the more or less local conditions of the experiment and experimentation itself are ultimately significant, and, moreover, now unlike in quantum mechanics, there are no reliable algorithms for estimating our chances. Our historical knowledge, gained from previous experience, may help us in shaping our expectations and making our estimates, but it only helps sometimes, and our estimates are nearly always probabilistic and ultimately uncertain. In particular, long-term histories define little and never determine the ontology of events, which ontology is discontinuous, eruptive with respect to any given history, although continuous with respect to other local histories.

Beckett's *Endgame* is an allegory of these conditions. There is no point in trying to finally understand the endgame of the play, since it is impossible, insofar as both randomness and coherence, an order that is only tentative and correlation-like in Beckett, are the products of that which is ultimately unthinkable. As noted earlier, Bohr replied to Harald Høffding's question "Where can the photon be said to be?" with "To be, to be, what does it mean to be?" (cited in Wheeler 1998, p. 131). Both of these questions are still unanswered and, in Bohr's view, they are indeed unanswerable: we don't know, we cannot know, or even think what this meaning can possibly be. Feynman's statement "Nobody understands quantum mechanics" is invoked often, and not infrequently as a repudiation of quantum mechanics, and as reflecting the fact that an alternative theory of quantum phenomena is desirable (Feynman 1965, p. 129). (Feynman's own position was more ambivalent and had undergone changes over the years.) However, that the ultimate behavior of quantum objects cannot be understood is part of Bohr's *understanding* of quantum phenomena and quantum mechanics, an epistemology that, Beckett tells us, extends beyond quantum physics, to the epistemology of life, as Bohr indeed suggests as well, as in the passage from "Unity of Knowledge" cited above (Bohr 1954, PWNB 2, p. 77). As noted from the outset, the very names "quantum," "object," and "process" are provisional: no concept associated with these names, or with any other names (names and concepts usually transferred from classical models) is rigorously applicable to quantum objects and processes, which are, thus, unnamable. Shelley spoke of the domain that he saw as allegorically defining the ultimate nature of light as "the realm without a name"

(*The Triumph of Life*, 1. 396). This unnamable makes possible certain physical effects, which are in turn responsible for certain nominal effects. *The Unnamable* is the title of Beckett's novel closing the Malloy trilogy, and one could give numerous quotations from it to illustrate the novel's quantum-like epistemology (Beckett 2009). By the same token, the novel and other works of Beckett also tell us that there may never be a master game, an *end* endgame. Instead, the game of probability, which Kant forbade in philosophy, even under the assumption of underlying causality, and moreover, probability without causality, becomes open and is interminable, and not only in practical matters, but against Kant's objection, as the defining game of philosophy (Kant 1997, pp. 384, 589, 661–662). This may not be what can be called classical philosophy, the philosophy based in reality and causality, wants. But, both Bohr and Beckett tell us, it is not the only philosophy, and it is not certain that this philosophy is always our best philosophical bet, in physics and beyond.

Chapter 10
1954–1962. "The Unity of Knowledge": New Harmonies

Bohr's ultimate interpretation of quantum phenomena and quantum mechanics, discussed in Chap. 9, was, I argue, essentially in place by the late 1930s. His 1949 "Discussion with Einstein on Epistemological Problems in Atomic Physics" reprised this interpretation, via his exchanges with Einstein, although without reflecting on changes in his thinking, in part under the impact of these exchanges. Instead, as also explained in Chap. 9, Bohr's earlier arguments are presented there through the optics of his later views. Einstein, too, had not proposed significantly new arguments after his 1936 "Physics and Reality," which refined EPR's argument, although he continued to offer analogous arguments concerning the EPR-type experiments into the 1940s. As explained in Chap. 8, his 1949 commentary on Bohr in the Schilpp volume essentially refers to Bohr's 1935 reply to EPR, which, as I noted, Einstein misread (Einstein 1949b, pp. 681–682). Nothing of substance was said there on Bohr's subsequent thinking, especially Bohr's ultimate epistemology presented in "Discussion with Einstein," published in the same volume. Thus, while continuing to shape a broader debate concerning quantum mechanics (this is still the case), substantively the Bohr–Einstein debate was over by around 1950 and in its essential features even earlier, and Bohr's and Einstein's views themselves were pretty much settled.

Bohr's *communications* on complementarity and the epistemology of quantum theory continued until his death in 1962. However, these communications (often occasioned by invited lectures on subjects beyond physics) do not develop substantively new arguments or offer significant refinements of previous arguments, and often repeat, with only minor changes and sometimes verbatim, his earlier formulations. One does find in these later works, beginning, roughly, with his 1954 "Unity of Knowledge," an additional degree of emphasis on "objectivity" and, sometimes, "causality." This emphasis led some commentators to argue that, contrary to my contention, there is a significant change in Bohr's views at this stage of his thinking, a change sometimes viewed positively (as a rapprochement with more classical epistemological views) and, on occasion, negatively (as a departure from his more radical earlier thinking and even as a "concession" to his Soviet colleagues [e.g., Gieser 2005, pp. 133-134]).

A. Plotnitsky, *Niels Bohr and Complementarity*, SpringerBriefs in Physics, DOI: 10.1007/978-1-4614-4517-3_10, © Arkady Plotnitsky 2013

Although an appeal to both *terms* is found in these later works, in contrast to his earlier essays, it does not appear to me to represent a substantive change in Bohr's argumentation itself. As indicated in Chap. 9, Bohr's meaning of these terms is in accord with his previous argumentation. "Unity of Knowledge" and other late essays define "objective" in terms of "unambiguous" communication, stressed by Bohr throughout his writings, especially in the wake of the EPR experiment. Bohr says: "Every scientist … is constantly confronted with the problem of objective description of experience by which we mean unambiguous communication" (Bohr 1954, *PWNB* 2, p. 67). Bohr amplifies this point, in connection with his concept of phenomenon, in "Quantum Physics and Philosophy: Causality and Complementarity" (1958), in speaking of "reserving the word 'phenomenon' solely for reference to unambiguously communicable information, in the account of which the word 'measurement' is used in its plain meaning of standardized comparison" (Bohr 1958, *PWNB* 3, p. 6). One might also recall that in Bohr's reply to EPR, whose criterion of reality contains, Bohr contends, "essential ambiguity," the concept of complementarity is argued to enable a disambiguation of this ambiguity, that is, an unambiguous attribution of all physical properties considered in quantum theory, and unambiguous communication of our experimental and theoretical findings. This disambiguation is a crucial aspect of complementarity, especially significant in Bohr's reply to EPR, and one of the main reasons for introducing it, and, as will be seen below, it is part of the genealogy of the concept.

The situation is similar as concerns Bohr's appeal to causality, in effect linked to an unambiguous communication, via the concept of verification. As Bohr says: "the emphasis on permanent recordings under well-defined experimental conditions [provides] the basis for a consistent interpretation of the quantum formalism correspond[ing] to the presupposition, *implicit in the classical physical account*, that every step of the *causal* sequence of events in principle allows for verification" (Bohr 1958, *PWNB* 3, p. 6; emphasis added). One should not confuse Bohr's use of "causal" here with ascribing causality to the individual quantum processes themselves. Bohr only refers to the verification of what is observed in measuring instruments. As Bohr stressed earlier in this article and on other occasions, any such verification is bound to be statistical, because no quantum experiment could be guaranteed to be repeated exactly. There is, thus, no question of applying causality to the behavior of quantum objects. The causal sequence in question only refers to an unambiguous communication and repetition of the conditions of measuring instruments: the same conditions will lead to the same statistical distribution in the recordings of the outcomes of thus repeated experiments. The article also makes clear that Bohr's ultimate epistemology remains fully in place, specifically insofar as quantum objects are placed outside of physical description and even beyond the reach of thought, which, again, makes the absence of causality at the quantum level automatic.

Thus, there appears to be no reason to associate Bohr's later arguments with any return to a more classical epistemology or with relinquishing any of his radical views. If anything, Bohr's later essays suggest a different view of what in fact defines "the classical physical account" as properly scientific, a view, however, also in place at least already in his reply to EPR. In this view, what makes a classical physical account, or any scientific account, properly scientific is the possibility of unambiguous verification and communication of experimental and theoretical findings, rather than the epistemology of classical physics, preferable and even imperative as this epistemology may be to some. It is sometimes argued that this, essentially realist, epistemology is necessary to ensure scientific objectivity and unambiguous communication. This is true as concerns objectivity, but only if the latter concept is defined in accordance with realism, in this case, amounting to the demand that a proper quantum theory should provide a description, even if idealized, of the properties and behavior of quantum objects themselves. This is not true, however, which is Bohr's point, as concerns unambiguous communication or the corresponding concept of objectivity. Part of Bohr's argument is that his epistemology, in conjunction with complementarity, "provides," to return to Bohr's formulation in his reply to EPR, "room for new physical laws the coexistence of which might at first sight appear irreconcilable with the basic principles of science" (Bohr 1935, p. 700). In this regard, it also enables the cohesiveness or "unity" of scientific knowledge in classical and quantum physics, or relativity.

The epistemological comparison of quantum theory and relativity (against the background of classical physics) is a persistent theme of Bohr's writings. In this respect, too, there is a kind of progression in these writings toward an understanding of relativity, especially general relativity, in terms of its greater proximity to quantum mechanics, vis-à-vis classical physics. It is not that key epistemological differences between both theories are not given proper attention. Bohr continues to stress these differences, especially on the account of causality: relativity, both special and general, is a causal theory, while quantum mechanics is not. The question of reality in both theories is more complicated, for the reason to be explained presently. Also, as we have seen, Bohr always had in mind parallels between both theories, such as that between the role of c (the speed of light in the vacuum) and Planck's constant h in establishing the break with classical physics in, respectively, relativity and quantum theory. Nevertheless, the affinities between quantum mechanics and relativity are given more emphasis in Bohr's works beginning with his reply to EPR, especially on the account of the role of measuring instruments in both theories. As Bohr writes in "Discussion with Einstein":

> Notwithstanding all differences between the physical problems which have given rise to the development of relativity theory and quantum theory, respectively, a comparison of purely logical aspects of relativistic and complementary argumentation reveals striking similarities as regards the renunciation of the absolute significance of conventional physical attributes of objects. Also, the neglect of the atomic constitution of the measuring instruments themselves, in the account of actual experience, is equally characteristic of the application of relativity and quantum theory. Thus, the smallness of the quantum of action compared with the actions involved in usual experience, including the arranging and handling of physical apparatus, is as essential in atomic physics as is the enormous number of atoms composing the world in the general theory of relativity which, as is often pointed out, demands that dimensions of apparatus for measuring angles can be made small compared with the radius of curvature of space. (Bohr 1949, *PWNB* 2, p. 64; also Bohr 1935, pp. 701–702)

Thus, it is the role of measuring instruments that assures the logical structure of relativity theory, special, or general. In relativity and quantum mechanics alike, the mathematical formalism of the theory coherently covers the situation of measurement, as Bohr notes in closing his reply to EPR (Bohr 1935, pp. 701–702). Bohr must have felt considerable frustration (there is evidence he did) that he failed to successfully convey to Einstein this significance of measuring instruments in the constitution of quantum phenomena, which Einstein clearly bypassed in his comments on Bohr's reply and throughout his arguments. Indeed, one of the reasons for Bohr's emphasis on this point might have been his aim to make his argumentation more convincing to Einstein. Einstein's position proved to be unshakable, however, and after a decade of the debate with Bohr, Einstein, again, only admitted that the kind of view adopted by Bohr is "logically possible without contradiction," while remaining "so very contrary to [Einstein's] scientific instinct" (Einstein 1936, p. 375). As "Discussion with Einstein" makes apparent, the unshakable nature of Einstein's position must have become clear to Bohr.

However, Bohr's argument expressed in the passage cited above has an independent significance not only in reflecting Bohr's views under discussion but also in its own right, and Einstein was not unaware of some of the problems involved. The problem of the physical constitution of the measuring instruments, rods, and clocks, in special and then general relativity troubled Einstein all his life, also in relation to the relationships between geometry and physics in relativity, as against Heisenberg's "algebraic" method in quantum mechanics (e.g., Einstein 1921, Einstein 1936, p. 378). I can, however, only mention this subject, without addressing it any further, given that Bohr's late thinking is my main focus here.[1] Bohr amplifies his argument in "Quantum Physics and Philosophy," from a related but different angle. He says:

> Notwithstanding all difference in the typical situations to which the notions of relativity and complementarity apply, they present in epistemological respects far-reaching similarities. Indeed, in both cases we are concerned with the exploration of harmonies which cannot be comprehended in the pictorial conceptions adapted to the account of more limited fields of physical experience. Still, the decisive point is that in neither case does the appropriate widening of our conceptual framework imply any appeal to the observing subject, which would hinder unambiguous communication of experience. In relativistic argumentation, such objectivity is secured by due regard to the dependence of the phenomena on the reference frame of the observer, while in complementary description all subjectivity is avoided by proper attention to the circumstances [the conditions of the measuring instruments] required for the well-defined use of elementary physical concepts. (Bohr 1958, *PWNB* 3, pp. 6–7)

Stressing that these epistemological complexities do not in any way hinder "unambiguous communication of experience", and hence "objectivity" (in Bohr's sense) is, again, characteristic of his later articles. However, as I argue and as this elaboration confirms as well, this emphasis in no way implies relinquishing any of Bohr's epistemology, but instead points toward "new harmonies which cannot be

---

[1] See (Brown and Pooley 2001) and (Heisenberg 1989, pp. 82–83).

comprehended in the pictorial conceptions" adapted to and, one might add (Bohr make this point elsewhere), arising from our phenomenal intuition [*Anschaulichkeint*]. Relativity and then, more radically, quantum mechanics revealed the limits of the applicability of such conceptions.

These difficulties become apparent early on, for example, if one tries to assign the classical spatial or temporal properties to a moving photon in special relativity. If it was possible to put a clock on a photon, it would stand still, while at the same time the photon would be found in all locations of its trajectory at once within its own spatial frame of reference. Rigorously, this means that the very concepts of clock and measuring rod lose their meaning in the frame of reference of a moving photon (there are no other photons), as does the concept of frame of reference. This situation reflects the fact that a photon is ultimately a quantum object and hence its relativistic treatment is an idealization, which is also causal in character. Nevertheless, Bohr's argument concerning affinities between quantum mechanics and relativity, especially as defined by a constitutive role of measuring instruments in both, remains in place. The problems of both causality and realism become more formidable in general relativity, even leaving aside the role of quantum theory there, for example, in considering the Hawking radiation of black holes. We might expect even greater, perhaps as yet unimaginable, complexities in the case of quantum gravity, although, as noted earlier, a return, hoped for by Einstein, to a more realist approach is not inconceivable either.

The choice of the word "harmonies" in Bohr's passage cited above is of some interest, and perhaps echoes or, in any event, is (to use another musical term) *in accord* with Einstein's 1949 description, cited in Chap. 2, of Bohr's 1913 atomic theory as "the highest musicality in the sphere of thought," Bohr's first exploration of these new harmonies. This description was clearly known to Bohr by the time he wrote this passage. What interests me most now, however, is Bohr's exploration of the harmonies between relativity and quantum mechanics. Harmonies (plural) may be a better word than unity, since these theories are hardly unified, physically, mathematically, or philosophically, or in any event not yet, and we do not know how they will be or even might possibly be. These harmonies are even richer, physically, mathematically, and philosophically, than Bohr suggests here, crucial as his points are, and they have a long and complex genealogy, which cannot be addressed here in detail. I would like, however, to address one particular juncture of this genealogy, which gives us a different and yet explored perspective on the history of physics, and of the relationships among mathematics, physics, and philosophy. This juncture brings me, in closing this book, to the earliest known evidence of complementarity, as recollected by Bohr in 1962 in his last interview with T. S. Kuhn, L. Rosenfeld, E. Rudinger, and A. Petersen, the last session of which took place literally on the eve of Bohr's death. I shall only be concerned with one point of the final session of this interview, where Bohr reflects on one of the key points in the origin of complementarity. As, I hope, this study has made clear, it is not possible to speak of a single such point, if we think of complementarity as the concept that defines Bohr's understanding of quantum phenomena and quantum mechanics because different components of this *multicomponent* concept emerged at different times and in response to different problems.

Before I discuss Bohr's reflection on his earlier thinking, I would like to note that, while Bohr's last interview sheds light on the history of his thinking and has an appeal or even aura of the last statement, it does not, in my view, add significantly to our understanding of this thinking, vis-à-vis his works considered here. In some respects, this interview is even disappointing, for example, as concerns Bohr's account of his exchange with EPR. It sharply contrasts with his writings on the subject considered in this study and sounds uncharacteristically defensive, almost angry, perhaps reflecting his frustration with his failure to successfully convey his argument to Einstein. It certainly does not do justice to the contribution of EPR's article, which is unquestionable, notwithstanding its problems, addressed (in a very different tone) by Bohr in his reply. On the other hand, it is Bohr's account of his first encounter with the situation, in this case in psychology, which suggests thinking in terms of complementarity, is revealing and significant.

We recall that in the Como lecture Bohr speaks of the "'inherent irrationality'" at the core of the quantum postulate (Bohr 1927, *PWNB* 1, p. 54). For the reasons explained earlier, he abandoned this way of speaking of quantum phenomena. As I argued in Chap. 4, however, his point itself was pertinent and philosophically profound. The idea of quantum "irrationality," in the direct sense of being beyond our thinking, had continued to shape his interpretation of quantum phenomena and quantum mechanics, a *rational* theory responding to quantum *irrationality*, from Heisenberg on. Bohr specifically referred to Heisenberg's discovery as that of a rational quantum mechanics (Bohr 1925, *PWNB* 1, p. 48). As I also explained in Chap. 4, Bohr was likely to have had in mind a parallel with the discovery of irrational numbers, such as $\sqrt{2}$ (or what we now thus designate), in ancient Greek mathematics, and then, in a more subtle way, with imaginary and complex numbers, on which the mathematical formalism of quantum mechanics essentially depends. He appears to have thought of complex numbers as "irrational" in the epistemological sense (although in mathematics the term "irrational numbers" is reserved for real irrationals and is not, generally, used in relation to complex numbers). That is, complex numbers, beginning with i, $\sqrt{-1}$, are irrational insofar as, just as regular (real) irrational numbers, they cannot be represented as a fraction of two integers. Also, in both cases we deal with the solutions of polynomial equations that do not belong to the field in which the equations themselves are defined, as in the case of $x^2 + 1 = 0$, the solutions of which are $\pm\, i$. For these reasons, the epistemological or even mathematical status of complex numbers was under debate for a long time, which is why they were called "*imaginary* numbers".

In any event, at stake is the impossibility of representing or even conceiving of something that nevertheless essentially related to a given, *rationally* formulated, theoretical framework, the situation taken to it, is at least thus far, the ultimate level in the case of the "algebra" of quantum mechanics. The solution of its equations or, more accurately, what these solutions ultimately relate to, that is, quantum objects is beyond the representational capacity of quantum mechanics, and even beyond the reach of thought altogether. This relation is, again, highly indirect and is possible only via probabilistic predictions concerning what is observed in measuring instruments impacted by quantum objects. Thus, while both the incommensurability of Greek mathematics and, differently, complex numbers could be represented geometrically, quantum objects equally defy algebraic and geometrical representation. It is true that in quantum mechanics, we deal with physical and not mathematical entities. However, at stake is also a difference from classical physics or (with the qualifications given above) relativity, whether mathematics functions descriptively. For the ancient Greeks, too, numbers or geometrical entities would describe the ultimate reality, material or spiritual, of the world. The "irrationality" inherent in the quantum-mechanical situation is made more intriguing because the quantum-mechanical formalism irreducibly depends on complex numbers, and relate, probabilistically, to the real (technically, rational) numbers obtained in measurement by means of additional *ad hoc* procedures. As Bohr said, the "symbols [used in the quantum-mechanical formalism], as indicated already by the use of imaginary numbers, are not susceptible to pictorial representation" (Bohr 1972–1999, v. 7, p. 314). Even if they were, it would not ultimately matter, because the formalism does not describe the behavior of quantum objects, although one might surmise that because it does not do so, this formalism is more likely to be grounded in complex than real numbers.

Intriguingly, these various facets of the irrational, and of rationally handling the irrational, come together in Bohr's final interview, in the remarks I want to consider and in the picture that he drew on the blackboard in the course of this interview (Fig. 10.1). First, Bohr reflected on the time before his interest in physics took over:

At that time I really thought to write something about philosophy, and that was about this analogy with multivalued functions. I felt that the various problems in psychology—which were called big philosophical problems, of the free will and such things—that one could really reduce them when one considered how one really went about them, and that was done on the analogy to multivalued functions. If you have square root of x, then you have two values. If you have a logarithm, you have even more. And the point is that if you try to say you have now two values, let us say of square root, then you can walk around in the plane, because, if you are in one point, you take one value, and there will be at the next point a value which is very far from it and one which is very close to it. If you, therefore, work in a continuous way, then you—I am saying this a little badly, but it does not matter—then you can connect the value of such a function in a continuous way. But then it depends what you do. If in these functions, as the logarithm or the square root, they have a singular value at the origin, then if you go round from one point and go in a closed orbit and it does not go round the origin, you come back to the same [value]. That is, of course, the discovery of Cauchy. But when you go round the origin, then you come over to the other [value of the] function, and that is then a very nice way to do it, as Dirichlet [Riemann], of having a surface in several sheets and connect them in such a way that you just have the different values of the function on the different sheets. And the nice thing about it is that you use one word for the function, $f(z)$. Now, the point is, what is the analogy? The analogy is that you say that the idea of yourself is singular in our consciousness—do you think it works; am I doing it sufficiently loud? Then you find—now it is really a formal way—that if you bring this idea in, then you leave a definite level of objectivity or subjectivity. For instance, when you have to do with the logarithm, then you can go around; you can change the function as much as you like; you can change it by $2\pi i$;

Fig. 10.1 The picture drawn by Bohr during his final interview

when you go one time round a singular point. But then you surely, in order to have it
properly and be able to draw conclusions from it, will have to go all the way back again in
order to be sure that the point is what you started on.—Now I am saying it a little badly,

but I will go on.—That is then the general scheme, and I felt so strongly that it was illuminating for the question of the free will, because if you go round, you speak about something else, unless you go really back again [the way you came]. That was the general scheme, you see. (Bohr 1962, Session 5)

The account thus represents the inaugural instance of the relationships between complementarity and psychology, an instance preceding quantum theory, but clearly affecting Bohr's subsequent extensions of complementarity to psychology. These aspects of Bohr's remarks have been noted by commentators. Here, I would like to reflect on a different aspect of the connections between Riemann's mathematics and Bohr's thinking and quantum theory, and twentieth-century physics, in general. Bohr, mistakenly, speaks of Gustaf Lejeune Dirichlet, Riemann's senior colleague at Göttingen and a great mathematician in his own right. Bohr clearly refers, however, to Riemann's concepts, with which he appears to have been familiar from early on, in part via his brother Harald Bohr, a major mathematician, whose dissertation was related to Riemann's work (on a different subject). Harald Bohr's main mathematical contributions were, however, in the field of functional analysis, which also deals with the mathematics (that of Hilbert spaces) used in quantum mechanics, a mathematics that has a Riemannian genealogy as well. Harald spent some time in the Hilbert circle in Göttingen, where his work was highly regarded. Later, he became a director of the Mathematical Institute, just across the street from Bohr's Physics Institute, in Copenhagen. Thus, Bohr, who talked to Harald daily, always had a ready expert source on the mathematics of quantum mechanics.

Now, at the top of the picture drawn by Bohr one finds a general formula for the functions of complex variables and a sketch of two separate complex-number sheets of a Riemann surface. Riemann introduces the idea of such a surface in dealing with the functions of complex variables to be able to have a single rather than a double (and hence ambiguous) meaning corresponding to each value argument. This makes it possible to treat such functions as regular mathematical functions, defined in mathematics so that only a single value of the function corresponds to each value of the argument. Thus, $y = \sqrt{x}$ is not a proper function, since for each x there are two values of y (say, if $x = 4$, y may be either 2 or -2). If we replace, the complex plane with a Riemann surface (corresponding to a given function) this difficulty disappear, in a kind of proto-complementary fashion. Just as complementarity, Riemann's concept enables a form of disambiguation, a concept the value of which Bohr appears to have initially appreciated in the case of psychological processes, but eventually transferred to quantum physics. In 1910, when he began his work on quantum theory, Bohr described his emotional state in complementary terms in a letter to Harald: "Emotions, like cognitions, must be arranged in planes that cannot be compared" (Letter to Harald Bohr, 26 June, 1910, Bohr 1972–1999, v. 1, p. 513; translation modified following D. Favrholdt, Bohr 1972–1999, v. 10, p. xxix). Whether this idea is ultimately workable in psychology or not (an open question, as noted in Chap. 9), it eventually worked in quantum theory, although it took 17 years and quantum mechanics to convert it into a properly quantum-theoretical concept of complementarity.

While it is true that Bohr's comments refer to his earlier thinking, the role of complex numbers in the quantum-mechanical formalism must have been on his mind at the time of the interview. Below on the black board there is a sketch of "Einstein's box" involved in Einstein's argument, discussed in Chap. 9, on which Bohr comments later in the interview. Bohr's analysis of the experiment embodies the essence of Bohr's quantum-mechanical thinking, especially if seen, as it was by Bohr in "Discussion with Einstein" and in this interview, through the optics of his analysis of the EPR experiment and his ultimate epistemology.

Thus, Bohr's sketch represents the lifelong trajectory of his thinking on complementarity and quantum theory, and the character of physics as a mathematical science, and indeed on the nature of human thinking and knowledge in general. Right underneath the formula for the functions of complex variables, Bohr writes $\sqrt{2}$, or at least something that suspiciously looks like 2, since it may also be z, customarily used to designate complex variables ($z = x + iy$), as it is by Bohr in the general formula above, where, however, z looks very different in his handwriting. If this is $\sqrt{2}$, it might have been used as both a general symbol of irrationality and as the opening move in a long history of the actual concept of the irrational, as the beyond thought, from ancient Greek mathematics to quantum mechanics, via modern algebra and (these fields are connected in turn) the theory of the functions of complex variables or, as it is known now, complex analysis. Although I suspect that Bohr has this history in mind even if it was $\sqrt{z}$ rather than $\sqrt{2}$, I shall not insist on this in the absence of firmer evidence to support it. In any event, the connections between Bohr's thinking and Riemann's theory could be ascertained. I would like now to comment on these connections and their implications for our understanding not only of Bohr's thinking but also of the history of quantum mechanics and twentieth-century physics in general.

I recall first that Riemannian geometry, defined by his concept of manifoldness (roughly a space that is Euclidean locally, or more accurately, infinitesimally, but in general not globally), is the basis of Einstein's general relativity theory, which was used by Bohr, against Einstein, in Bohr's analysis of Einstein's photon-box experiment. Special relativity, too, may be recast in a Riemannian form, as was done, in 1908, by Hermann Minkowski, who introduced the concept of "space-time" as part of this recasting. Riemann, who had an abiding interest in and made major contributions to physics, made intriguing, albeit tentative, steps toward unifying gravitation and light, and he came close to some of Einstein's ideas, although not those of general relativity, in which gravity defines the geometry of a physical space by curving it.

Thus, Riemann's mathematical thinking has connections to both Einstein's general relativity and Bohr's concept of complementarity. One might object that bringing the Riemannian aspects of these theories together in this way is artificial. First of all, we deal with two different mathematical theories: (real) differential geometry and the theory of the functions of complex variables. Second, while Einstein expressly uses Riemann's geometry as the (descriptive) mathematical formalism of his theory, in which the curvature of space is defined by gravity, Bohr borrows an essentially philosophical point of disambiguation found in Riemann's theory of the functions of complex variables. Both of these points are true, and it is not my aim to claim *immediate* connections along these lines between relativity and complementarity. However, as concerns the first point, it is well known, Riemann's thinking concerning his geometry was an extension of his thinking concerning Riemann surfaces, which are the first example of manifolds (in two dimensions) in Riemann's sense, especially as concerns the topological aspects of both concepts (e.g., Ferreirós 2006). The invention of Riemann surfaces was one of the crucial events in the history of topology as a mathematical discipline.

As concerns the second point, I would like to place it in a broader perspective, defined by the intersection of mathematics, physics, and philosophy, "a magic triangle," as it has been called.[2] It can be shown that Riemann's famous Habilitation lecture "On Hypotheses That Lie at the Foundations of Geometry" [*Ueber die Hypothesen, welche der Geometrie zu Grunde liegen*] (Riemann 1854), which introduced Riemannian geometry, is as much an exercise in foundational thinking in *physics* as in mathematics. It is also an exercise in foundational thinking on the relationships between mathematics and physics. Riemann's title refers even more to physics than to mathematics. At stake are foundations of geometry, to be sure, and the lecture offers extraordinary new mathematics. However, as most of his contemporaries, Riemann saw geometry as a science of physical space; and the hypotheses in question concern physical space, and are to be tested physically, experimentally. Moreover, Riemann expressly brings into his analysis important physical considerations, and offers profound philosophical reflections concerning the relationships between mathematics and physics. This is not surprising. Riemann's interests in physics are well known: during his early years in Göttingen, eventually the birth place of quantum mechanics, he worked as an assistant to Wilhelm Weber. Riemann wrote several papers and extensive notes on various aspects of physics, in particular, on electromagnetism. The connections between physics and mathematics were close in Riemann's time. They were important for the work of his teacher Gauss, who pursued important research in electromagnetism, and collaborated with Weber on the invention of one of the first telegraphs. Gauss's work on the geometry of curved surfaces, specifically the concept of Gauss's curvature, was decisive for Riemann's work on geometry from the mathematical side. Around the same time, however, the close connections between mathematics and physics were also beginning to break with the emergence of such fields as the algebraic number theory (to which Gauss made major contributions) and more abstract forms of analysis and geometry, such as projective geometry. These developments eventually made mathematics independent from physics. This independence is characteristic of what is now called "modern mathematics," which emerged, roughly, around Riemann's time.[3]

Some of Riemann's work, including his work on geometry, had contributed to this independence. Riemann's lecture curiously reflects this situation by applying

---

[2] I shall only offer a sketch here, based on a more extensive treatment in (Plotnitsky 2012).

[3] On "modern mathematics," see (Gray 2008).

the term "space" strictly to physical space and, it appears, the term "geometry" to the study of continuous manifolds of any dimensions as a purely mathematical object. There is no paradox here. First, while Riemann's mathematics was responsible for several key theories and ways of thinking that led to the emergence of modern mathematics as independent of physics, Riemann himself was not thinking in terms of separating mathematics from physics, either in most of this work or in his philosophy of mathematics. In this respect, he was much closer to Gauss than to such of his younger contemporaries as R. Dedekind, L. Kronecker, and G. Cantor, or such earlier figures as N. H. Abel and E. Galois. These mathematicians were also responsible for the rise of modern mathematics, but they did not appear to have had much interest in physics. Riemann's thought is sometimes seen in terms of a "magic triangle" of mathematics, physics, and philosophy, just as is later that of Einstein and Weyl, arguably the figure closest to Riemann in this regard, for whom Riemann was a major influence and inspiration.[4] Second, one can also see modern mathematics not only as breaking, by its highly abstract character, previously established connections between mathematics and physics, but also as creating a possibility of new connections, and new *types* of connections, between them. Such different connections are found, for example, in general relativity, which broke, via Riemannian geometry, the previous link between physics and Euclidean geometry.

Now, arguably the single most important epistemological point of Bohr's work on quantum mechanics and of this study is that the situation takes a more radical form in quantum mechanics, beginning with Heisenberg's invention of the theory. In relativity, mathematics is still used, just as it is in classical physics, to develop *idealized* descriptive models of the physical objects considered and to make predictions, which relate to what is *actually* observed, on the basis of the idealized descriptions these models provide. As explained above, relativity does bring with it certain complexities (vis-à-vis classical physics) into this situation, and one finds views of relativity similar to the present view of quantum mechanics. While it is, again, not something that would have appealed to Einstein or appeals to most even now; this type of view would reinforce my main point at the moment by bringing relativity closer to quantum mechanics.

What I would especially like to stress here is that Heisenberg used and in part reinvented modern, more abstract mathematics (that of infinite-dimensional matrix algebra) for the purposes of these predictions. As happened, this mathematics, de facto the mathematics of the infinite-dimensional Hilbert spaces, has Riemannian genealogy as well, specifically in Riemann's Habilitation lecture, where Riemann suggested, arguably for the first time ever, the idea of metrical spaces of an infinite number of dimensions, as Hilbert spaces do in the case of continuous variables (such as position and momentum). In recent years, there have also emerged intriguing connections along these lines between quantum field theory and

---

[4] The phrase "the magic triangle" is due to S. Ron (cited in Ferreirós 2006, p. 67). On Weyl, see (Wheeler 1994).

Riemann's work on the distribution of primes and his famous hypothesis concerning the ζ-function. The mathematics (such as that of the so-called Calabi-Yau spaces) used in current, still hypothetical, theories, such as string and brane theories, that aim to bring together general relativity and quantum mechanics, are, too, an extension in Riemann's geometrical ideas, which, again, laid the foundation for twentieth-century topology and geometry, extensively used in these theories. These subjects, however, would take us too far afield. My main point is that Heisenberg took advantage of this abstraction to establish new, predictive (probabilistically), rather than descriptive use of mathematics in quantum theory.

His approach, as we have seen, compelled Bohr to speak of "the new epoch of mutual stimulation of mathematics and mechanics," via this new type of mutual relationship, enabled by a loss of the descriptive capacity of formalism (Bohr 1925, p. 51). Heisenberg's radical move was beyond what Riemann (whose position is closer to that of Einstein) ever suggested. I would argue, however, that Riemann's thinking prepares this type of move, as Bohr, I would also argue, realized. Complementarity brings yet another Riemannian connection to this situation, that defined by Riemann's proto-complementary idea of disambiguation of the functions of complex variables, which he coupled to Heisenberg's epistemology, eventually in turn brought by Bohr to its limit in his interpretation of quantum phenomena and quantum mechanics.

Whether Bohr himself thought along these lines, as I suspect he might have at least at the time of his interview in 1962, the connections between complementarity and Riemann's mathematics direct us to an extraordinarily rich history of thought, defined by the magic triangle of mathematics, physics, and philosophy. This history is barely explored as yet, even in considering Riemann, Einstein, and Bohr's thinking, itself a magic triangle of proper names, names of ways of thinking and as yet unsolved problems. In and beyond their thinking, we are, as Bohr envisioned, bound to discover new harmonies of thought and knowledge, and of the relationships among mathematics, physics, and philosophy, harmonies that reflect, to bring Einstein and Bohr together for the last time, "the highest musicality in the sphere of thought".

References

Archive for the history of quantum physics (University of Pittsburg). http://www.library.pitt.edu/libraries/special/asp/quantum.html.

Aspect, A., J. Dalibard, and G. Roger. 1982. Experimental test of Bell's inequalities using time-varying analyzers. *Physical Review Letters* 49: 1804.

Atmanspacher, H., and H. Primas, eds. 2009. *Recasting reality: Wolfgang Pauli philosophical ideas and contemporary science*. Berlin: Springer.

Beckett, S. 1981. *Endgame*. New York: Grove.

Beckett, S. 2008. In *Waiting for Godot*, ed. H. Bloom. New York: Chelsea House.

Beckett, S. 2009. *Three novels: Molloy, Malone Dies, and the unnamable*. New York: Grove.

Bell, J.S. 2004. *Speakable and unspeakable in quantum mechanics*. Cambridge: Cambridge University Press.

Beller, M. 1999. *Quantum dialogue: The making of a revolution*. Cambridge: Cambridge University Press.

Bertlmann, R.A. and A. Zeilinger, eds. 2002. *Quantum (un)speakables: From bell to quantum information*. Berlin: Springer.

Bohr, A., B.R. Mottelson, and O. Ulfbeck. 2004. The principles underlying quantum mechanics. *Foundations of Physics* 34: 405.

Bohr, N., *Niels Bohr archive*. College Park, MD: Copenhagen and American Institute of Physics.

Bohr, N. 1913. On the constitution of atoms and molecules (part 1). *Philosophical Magazine* 26: 1.

Bohr, N. 1925. Atomic theory and mechanics. In *Philosophical writings of Niels Bohr*, 3 vols, ed. N. Bohr, vol. 1, 25–51. Woodbridge, CN: Ox Bow Press, 1987.

Bohr, N. 1927. The quantum postulate and the recent development of atomic theory. In *Philosophical writings of Niels Bohr*, 3 vols, ed. N. Bohr, vol. 1, 52–91. Woodbridge, CN: Ox Bow Press, 1987.

Bohr, N. 1929a. The quantum of action and the description of nature. In *Philosophical writings of Niels Bohr*, 3 vols. ed. N. Bohr, vol. 1, 92–101. Woodbridge, CN: Ox Bow Press, 1987.

Bohr, N. 1929b. Introductory survey. In *Philosophical writings of Niels Bohr*, 3 vols, ed. N. Bohr, vol. 1, 1–24. Woodbridge, CN: Ox Bow Press, 1987.

Bohr, N. 1931. Space-time continuity and atomic physics. In *Niels Bohr: Collected works*, vol. 6, 361–370. Amsterdam: Elsevier, 1972–1996.

Bohr, N. 1935. Can quantum-mechanical description of physical reality be considered complete? *Physical Review* 48: 696.

Bohr, N. 1937. Causality and complementarity. In *The philosophical writings of Niels Bohr, volume 4: Causality and complementarity, supplementary papers*, eds. J. Faye and H.J. Folse, 83–91. Woodbridge, CT: Ox Bow Press, 1998.

A. Plotnitsky, *Niels Bohr and Complementarity*, SpringerBriefs in Physics,
DOI: 10.1007/978-1-4614-4517-3, © Arkady Plotnitsky 2013

Bohr, N. 1938. The causality problem in atomic physics. In *The philosophical writings of Niels Bohr, volume 4: Causality and complementarity, supplementary papers,* eds. J. Faye and H.J. Folse, 94–121. Woodbridge, CT: Ox Bow Press, 1998.

Bohr, N. 1948. On the notions of causality and complementarity. In *The philosophical writings of Niels Bohr, volume 4: Causality and complementarity, supplementary papers,* eds. J. Faye and H.J. Folse, 141–148. Woodbridge, CT: Ox Bow Press, 1998.

Bohr, N. 1949. Discussion with Einstein on epistemological problems in atomic physics. In *Philosophical writings of Niels Bohr,* 3 vols, ed. N. Bohr, vol. 2, 32–66. Woodbridge, CN: Ox Bow Press, 1987.

Bohr, N. 1954. Unity of knowledge. In *Philosophical writings of Niels Bohr,* 3 vols, vol. 2, 67–82. Woodbridge, CN: Ox Bow Press, 1987.

Bohr, N. 1956. Mathematics and natural philosophy. In *The philosophical writings of Niels Bohr, volume 4: Causality and complementarity, supplementary papers,* eds. J. Faye and H.J. Folse, 164–169. Woodbridge, CT: Ox Bow Press, 1998.

Bohr, N. 1958. Quantum physics and philosophy—causality and complementarity. In *Philosophical writings of Niels Bohr,* 3 vols, vol. 3, 1–7. Woodbridge, CN: Ox Bow Press, 1987.

Bohr, N. 1962. Interview with T. Kuhn, L. Rosenfeld, A. Petersen, and E. Rüdinger, 17 Nov 1962. In *Niels Bohr archive.* College Park, MD: Copenhagen and American Institute of Physics. http://www.aip.org/history/ohilist/4517_1.html

Bohr, N. 1972–1999. *Niels Bohr: Collected works,* 10 vols. Amsterdam: Elsevier, 1972–1996.

Bohr, N. 1987. *The philosophical writings of Niels Bohr,* 3 vols. Woodbridge, CT: Ox Bow Press, 1987.

Bohr, N. 1998. *Philosophical writings of Niels Bohr, volume 4: Causality and complementarity, supplementary papers,* eds. J. Faye and H.J. Folse. Woodbridge, CT: Ox Bow Press, 1998.

Bohr, N., and L. Rosenfeld. 1933. On the question of the measurability of electromagnetic field quantities. In *Quantum theory and measurement,* eds. J.A. Wheeler and W.H. Zurek, 479–522. Princeton, NJ: Princeton University Press, 1983.

Bohr, N., and L. Rosenfeld. 1950. Field and charge measurements in quantum electrodynamics. In *Quantum theory and measurement,* ed. J.A. Wheeler, and W.H. Zurek, 523–534. Princeton, NJ: Princeton University Press, 1983.

Bohr, N., H.A. Kramers, and J.C. Slater. 1924. The quantum theory of radiation. *Philosophical Magazine* 47: 785.

Born, M. 1926. Quantenmechanik der Stoßvorgänge. *Zeitschrift Für Physik* 38: 803.

Born, M. 2005. *The Einstein-Born letters* (trans: Born, I.). New York: Walker.

Born, M., W. Heisenberg, and P. Jordan. 1926. On quantum mechanics II. In *Sources of quantum mechanics,* ed. B.L. van der Warden, 321–385. New York: Dover, 1968.

Born, M., and P. Jordan. 1925. In *Sources in quantum mechanics,* ed. B.L. van der Warden, 277–306. New York: Dover, 1968

Brown, H.R., and O. Pooley. 2001. The origin of spacetime metric: Bell's 'Loretzian pedagogy' and its significance in general relativity. *Physics meets philosophy at the Planck scale: Contemporary theories of quantum gravity,* eds. C. Callender and N. Huggett, 256–272. Cambridge: Cambridge University Press, 2001.

Cao, T. Y., ed. 2004. *Conceptual Foundations of Quantum Field Theory.* Cambridge: Cambridge University Press

Cushing, J.T., and E. McMullin, eds. 1989. *Philosophical consequences of quantum theory: Reflections on Bell's theorem.* Notre Dame, IN: Notre Dame University Press.

Derrida, J. 1992. Before the law. In *Acts of literature,* ed. D. Attridge. New York: Routledge.

Dirac, P.A.M. 1925. The fundamental equations of quantum mechanics. In *Sources of quantum mechanics,* ed. B.L. van der Warden, 307–320. New York: Dover, 1968.

Dirac, P.A.M. 1927. The physical interpretation of the quantum dynamics. *Proceedings of Royal Society of London A* 113: 621.

Dirac, P.A.M. 1958. *The principles of quantum mechanics.* Oxford: Clarendon, rpt., 1995.

Dyson, F.J. 1949. The S-matrix in quantum electrodynamics. *Physical Review* 75: 1736.

Dyson, F.J. 2005. Hans Bethe and quantum electrodynamics. *Physics Today* 58: 48.

Einstein, A. 1921. Geometry and experience. In *Ideas and opinions*, 232–246. New York: Random House, 1988.

Einstein, A. 1925. "*Quantentheorie des einatomigen idealen Gases,*" *Sitz. ber. Preuss. Akad. Wiss. (Berlin)*, 3–14 (presented at the meeting of 29 Jan 1925).

Einstein, A. 1936. Physics and reality. *Journal of the Franklin Institute* 221: 349.

Einstein, A. 1948. Quantum mechanics and reality. *Dialectica* 2: 320–324; reprinted in English in M. Born, *The Born-Einstein letters* (trans: Born, I.), 168–173. New York: Walker, 2005.

Einstein, A. 1949a. *Autobiographical notes* (trans: Schilpp, P.A.). La Salle, IL: Open Court, 1991.

Einstein, A., 1949b. Remarks to the essays appearing in this collective volume. In *Albert Einstein: Philosopher-scientist*, ed. P.A. Schilpp, 663–688. New York: Tudor.

Einstein, A., B. Podolsky, and N. Rosen. 1935. Can quantum-mechanical description of physical reality be considered complete? In *Quantum theory and measurement*, ed. J.A. Wheeler and W.H. Zurek, 138–141. Princeton, NJ: Princeton University Press.

Ellis, J., and D. Amati, eds. 2000. *Quantum reflections*. Cambridge: Cambridge University Press.

Ferreirós, J. 2006. Riemann's *Habilitationsvortrag* at the crossroads of mathematics, physics, and philosophy. In *The architecture of modern mathematics: Essays in history and philosophy*, ed. J. Ferreirós, and J.J. Gray, 67–96. Oxford: Oxford University Press.

Feynman, R. 1965. *The character of physical law*. Cambridge, MA: MIT Press, rpt., 1994.

Feynman, R. 1985. *QED: The strange theory of light and matter*. Princeton, NJ: Princeton University Press.

Folse, H.J. 1985. *The philosophy of Niels Bohr: The framework of complementarity*. Amsterdam: North Holland.

Folse, H.J. 1987. Niels Bohr's concept of reality. In *Symposium on the foundations of modern physics 1987: The Copenhagen interpretation 60 years after the Como lecture*, eds. P. Lahti and P. Mittelstaedt, 161–180. Singapore: World Scientific.

Folse, H.J. 2002. Bohr's conception of the quantum-mechanical state of a system and its role in the framework of complementarity. In *Quantum theory: Reconsiderations of foundations*, ed. A. Khrennikov, 83–98. Växjö: Växjö University Press.

Gieser, S. 2005. The innermost kernel: depth psychology and quantum physics. Wolfgang Pauli's Dialogue with C. G. Jung. Berlin and Heidelberg: Springer.

Gomatam, R. 2007. Niels Bohr's interpretation and the Copenhagen interpretation—are the two incompatible. *Philosophy of Science* 74: 736.

Gottfried, K. 2000. Does quantum mechanics carry the seeds of its own destruction. In *Quantum reflections*, eds. J. Ellis and D. Amati, 165–185. Cambridge: Cambridge University Press.

Gray, J. 2008. *Plato's ghost: The modernist transformation of mathematics*. Princeton, NJ: Princeton University Press.

Greenberger, D.M., M.A. Horne, and A. Zeilinger. 1989. Going beyond Bell's theorem. In *Bell's theorem, quantum theory and Conceptions of the universe*, ed. M. Kafatos, 69–72. Dordrecht: Kluwer.

Greenberger, D.M., M.A. Horne, A. Shimony, and A. Zeilinger. 1990. Bell's theorem without inequalities. *American Journal of Physics* 58: 1131.

Hardy, L. 1993. Nonlocality for two particles without inequalities for almost all entangled states. *Foundations of Physics* 13: 1665.

Heidegger, M. 1967. *What is a thing?* (trans: W.B. Barton, Jr. and V. Deutsch). South Bend, IN: Gateway.

Heisenberg, W. 1925. Quantum-theoretical re-interpretation of kinematical and mechanical relations. In *Sources of quantum mechanics*, ed. B.L. Van der Waerden, 261–277. New York: Dover, 1968.

Heisenberg, W. 1927. The physical content of quantum kinematics and mechanics. In *Quantum theory and measurement*, eds. J.A. Wheeler and W.H. Zurek, 62–86. Princeton, NJ: Princeton University Press, 1983.

Heisenberg, W. 1930. *The physical principles of the quantum theory* (trans: K. Eckhart and F. C. Hoyt). New York: Dover, rpt. 1949.

Heisenberg, W. 1935, *"Ist eine deterministische Ergänzung der Quantenmechanik möglich?" Archive for the history of quantum physics* (microfilm 45, section 11), English translation by E. Crull and G. Bacciagaluppi, translation of: W. Heisenberg, "Ist eine deterministische Ergänzung der Quantenmechanik möglich?". http://philsci-archive.pitt.edu/8590/1/Heis1935_EPR_Final_translation.pdf

Heisenberg, W. 1962. *Physics and philosophy: The revolution in modern science*. New York: Harper & Row.

Heisenberg, W. 1967. Quantum theory and its interpretation. In *Niels Bohr: His life and work as seen by his friends and colleagues*, ed. S. Rozental, 94–108. Amsterdam: North-Holland.

Heisenberg, W. 1971. *Physics and beyond: Encounters and conversations*. London: G. Allen & Unwin.

Heisenberg, W. 1989. *Encounters with Einstein, and other essays on people, places, and particles*. Princeton, NJ: Princeton University Press.

Held, K. 2006. The Kochen-Specker theorem. In *Stanford encyclopedia of philosophy*. http://plato.stanford.edu/entries/kochen-specker/.

Howard, D. 2004. Who invented the 'Copenhagen interpretation'? A study in mythology. *Philosophy of Science* 71(5): 669.

Jaeger, G. 2007. *Quantum information: An overview*. New York: Springer.

James, W. 2007. *The principles of psychology*, vol. 1. New York: Cosimo Classics.

Kafatos, M., ed. 1989. *Bell's theorem, quantum theory and conceptions of the universe*. Dordrecht: Kluwer.

Kant, I. 1997. *Critique of pure reason* (trans: P. Guyer and A.W. Wood). Cambridge: Cambridge University Press.

Landau, L., and R. Peierls. 1931. Extension of the uncertainty principle to relativistic quantum theory. In *Quantum theory and measurement*, eds. J.A. Wheeler and W.H. Zurek, 465–479. Princeton, NJ: Princeton University Press.

Mehra, J. and H. Rechenberg. 2001. *The historical development of quantum theory*, 6 vols. Berlin: Springer.

Mermin, N.D. 1990. *Boojums all the way through*. Cambridge: Cambridge University Press.

Mermin, N.D. 1998. What is quantum mechanics trying to tell us? *American Journal of Physics* 66: 753.

Mermin, N.D. 2007. *Quantum computer science: An introduction*. Cambridge: Cambridge University Press.

Murdoch, D. 1987. *Niels Bohr's philosophy of physics*. Cambridge: Cambridge University Press.

Pais, A. 1982. *Subtle is the lord: The science and the life of Albert Einstein*. Oxford: Oxford University Press.

Pais, A. 1986. *Inward bound: Of matter and forces in the physical world*. Oxford: Oxford University Press.

Pais, A. 1991. *Niels Bohr's times, in physics, philosophy, and polity*. Oxford: Clarendon.

Pauli, W. 1994. *Writings on physics and philosophy*. Berlin: Springer.

Peres, A. 1993. *Quantum theory: Concepts and methods*. Dordrecht: Kluwer.

Petruccioli, S. 2011. Complementarity before uncertainty. *Archive for the History of Exact Sciences* 65: 591.

Petersen, A. 1985. The philosophy or Niels Bohr. In *Niels Bohr: A centenary volume*, eds. A.P. French, and P.J. Kennedy. Cambridge, MA: Harvard University Press.

Plotnitsky, A. 2002. *The knowable and the unknowable: Modern science, nonclassical thought, and the "two cultures"*. Ann Arbor, MI: University of Michigan Press.

Plotnitsky, A. 2006. *Reading Bohr: Physics and philosophy*. Dordrecht: Springer.

Plotnitsky, A. 2009. *Epistemology and probability: Bohr, Heisenberg, Schrödinger and the nature of quantum-theoretical thinking*. New York: Springer.

Plotnitsky, A. 2012. 'To be, To be, What Does It Mean to Be?': On quantum-like literary models. In *Foundations of probability of physics-6*, eds. M. D'Ariano et al., 463–487. Melville, NY: American Institute of Physics.

Randall, L. 2005. *Warped passages: Unraveling the mysteries of the universe's hidden dimensions*. New York: Harpers Collins.

Riemann, B. 1954. On the hypotheses that lie at the foundations of geometry. In *Beyond geometry: Classic papers from Riemann to Einstein*, ed. P. Pesic, 23–40. Mineola, NY: Dover.

Rosenfeld, L. 1963. Introduction. In *On the constitution of atoms and molecules. Papers of 1913 reprinted from the philosophical magazine*, ed. N. Bohr. Copehagen: Munksgaard; New York: W. A. Benjamin.

Rosenfeld, L. 1967. Niels Bohr in the thirties: Consolidation and extension of the conception of complementarity. In *Niels Bohr: His life and work as seen by his friends and colleagues*, ed. S. Rozental. Amsterdam: North Holland.

Schilpp, P. A., ed. 1949. *Albert Einstein: Philosopher-scientist*. New York: Tudor.

Schlosshauer, M. 2007. *Decoherence and the quantum-to-classical transition*. Heidelberg: Springer.

Schrödinger, E. 1935. The present situation in quantum mechanics. In *Quantum theory and measurement*, eds. J.A. Wheeler, and W.H. Zurek. Princeton, NJ: Princeton University Press.

Schweber, S.S. 1994. *QED and the men who made it: Dyson, Feynman, Schwinger, and Tomonaga*. Princeton, NJ: Princeton University Press.

Shimony, A. 2004. "Bell's theorem". In *Stanford encyclopedia of philosophy*. http://plato.stanford.edu/entries/bell-theorem/

Stapp, H.P. 2007. *Mindful universe: Quantum mechanics and participating observer*. Heidelberg: Springer.

Teller, P. 1995. *An interpretive introduction to quantum field theory*. Princeton, NJ: Princeton University Press.

Ulfbeck, O., and A. Bohr. 2001. Genuine fortuitousness: Where did that click come from? *Foundations of Physics* 31: 757.

Van der Waerden, B.L., ed. 1968. *Sources of quantum mechanics*. New York: Dover.

Von Neumann, J. 1932. *Mathematical foundations of quantum mechanics* (trans: R.T. Beyer). Princeton, NJ: Princeton University Press, rpt., 1983).

Weinberg, S. 2005. *The quantum theory of fields, volume 1: Foundations*. Cambridge: Cambridge University Press.

Weizsäcker, C.F. 1971. The copenhagen interpretation. In *Quantum theory and beyond*, ed. T. Bastin. Cambridge: Cambridge University Press.

Wheeler, J.A. 1983. Law without law. In *Quantum theory and measurement*, eds. J.A. Wheeler and W.H. Zurek. Princeton, NJ: Princeton University Press.

Wheeler, J.A. 1994. Foreword. In *The continuum: A critical examination of the foundation of analysis* (trans: S. Pollard and T. Bole), 9–14. New York: Dover, rpt., 1918.

Wheeler, J.A. 1998. *Geons, black holes, and quantum foam: A life in physics*. New York: W. W. Norton.

Wheeler, J.A., and W.H. Zurek (eds.). 1983. *Quantum theory and measurement*. Princeton, NJ: Princeton University Press.

Whitehead, A.N. 1929. *Process and reality: An essay on cosmology*. New York: Simon and Schuster, rpt., 1979.

Wilczek, F. 2005. In search of symmetry lost. *Nature* 423: 239.

Wittgenstein, L. 1924. *Tractatus Logico-Philosophicus* (trans: C.K. Ogden). London: Routledge, rpt., 1985.

Zeilinger, A., G. Weihs, T. Jennewein, and M. Aspelmeyer. 2005. Happy centenary, photon. *Nature* 433: 230.

Zurek, W.H. 2003. Decoherence, einselection and the quantum origin of the classical. *Review of Modern Physics* 75: 715.

Further Reading

Favrhold, D. 1992. *Niels Bohr's philosophical background*. Copenhagen: Det Kongelige Danske Videnskabernes Selskab.

Faye, J. 1991. *Niels Bohr: His heritage and legacy. An anti-realist view of quantum mechanics*. Dordrecht: Kluwer.

Faye, J., and H. Folse, eds. 1994. *Niels Bohr and contemporary philosophy*. Dordrecht: Kluwer.

Folse, H.J. 1985. *The philosophy of Niels Bohr: The framework of complementarity*. Amsterdam: North Holland.

French A.P., and P.J. Kennedy, eds. 1985. *Niels Bohr: A centenary volume*. Cambridge, MA: Harvard University Press.

Heisenberg, W. 1962. *Physics and philosophy: The revolution in modern science*. New York: Harper & Row.

Heisenberg, W. 1971. *Physics and beyond: Encounters and conversations*. London: G. Allen & Unwin.

Honner, J. 1987. *The description of nature: Niels Bohr and the philosophy of quantum physics*. Oxford: Clarendon, 1987.

Jammer, M. 1989. *The conceptual development of quantum mechanics: Interpretations of quantum mechanics in historical perspective*. Melville, NY: American Institute of Physics.

Murdoch, D. 1987. *Niels Bohr's philosophy of physics*. Cambridge: Cambridge University Press.

Pais, A. 1991. *Niels Bohr's times, in physics, philosophy, and polity*. Oxford: Clarendon.

Petruccioli, S. 2006. *Atoms, metaphors and paradoxes: Niels Bohr and the construction of new physics* (trans: I. McGilvray). Cambridge: Cambridge University Press.

Rozental, S. 1967. *Niels Bohr: His life and work as seen by his friends and colleagues*. Amsterdam: North-Holland.

Author Index

A
Abel, N.H., 178
Amati, D., 113n5, 183.
Aspect, A., 109, 181
Aspelmeyer, M., 185
Atmanspacher, H., 159n6, 181

B
Beckett, S., 160–165, 181
Bell, J. S., 107, 113n5, 123, 181
Beller, M., 27n1, 181
Bertlmann, R. A., 113n5, 118
Bethe, H., 97
Bohr, A., 82n6, 182, 185
Bohr, H., 175
Bohr, N.
Boltzmann, L., 160
Born, M., 2, 13, 29–33, 36–38, 81, 90, 92, 110, 130, 150, 155–157, 163, 182
Bothe, W., 3, 108
Brown, H. R., 170n1, 182

C
Cantor, G., 178
Cauchy, A-L., 173
Cayley, A., 137
Coleman, S., 161
Cushing, J. T., 113, 182

D
Dalibard, J., 181
Darwin, C., 160

De Broglie, L., 44, 56, 77n4, 81, 84
Dedekind, R., 178
Democritus, x, 148, 149
Derrida, J., 160n7, 182
Dirac, P. A. M., x, 2, 2n1, 8, 30, 35–38, 41, 49, 51, 54, 55, 84, 90–94, 98–105, 182
Dirichlet, G. L., 173–175
Dostoyevsky, F., 162
Dyson, F., 97–99, 182

E
Einstein, A.
Ellis, J., 113, 183

F
Faraday, M., 91
Favrhold, D., 19n1, 175, 186
Faye, J., 2n1, 181, 182, 186
Fermi, E., 91, 93
Ferreirós. J., 177, 178n4, 183
Feynman, R., 74n3, 77, 97, 140, 164, 183
Folse, H., 2n1, 148n5, 181–183, 186
French, A. P., 184, 186
Freud, S., 159n6

G
Galileo, 5, 19, 30, 37, 40, 158
Galois, E., 178
Gauss, K. F., 177, 178
Geiger, H., 3, 108, 156
Gieser, S., 159, 167, 183

A. Plotnitsky, *Niels Bohr and Complementarity*, SpringerBriefs in Physics, DOI: 10.1007/978-1-4614-4517-3, © Arkady Plotnitsky 2013

Subject Index

A. Plotnitsky, *Niels Bohr and Complementarity*, SpringerBriefs in Physics,
DOI: 10.1007/978-1-4614-4517-3, © Arkady Plotnitsky 2013